External Works and Drainage – A Practical Guide

External Works and Drainage – A Practical Guide

Phil Pitman

First Published 2001
By Taylor & Francis
2 Park Square, Milton Park, Abingdon, Oxon, OX14 4RN

Simultaneously published in the USA and Canada
By Taylor & Francis
270 Madison Ave, New York NY 10016

Taylor and Francis is an imprint of the Taylor and Francis Group

Transferred to Digital Printing 2006

Publisher's Note
This book has been prepared from camera-ready copy provided by the author.

British Library Cataloguing in Publication Data
A catalogue record for this book is available from the British Library

Library of Congress Cataloging in Publication Data
A catalog record for this book has been requested

ISBN 0-419-25760-8 (pbk)

Table of Contents

Preface

Welcome to the first edition of *External Works, Roads and Drainage – A Practical Guide*. In this first edition, I have tried to incorporate many of the different aspects of road, drainage and external works design.

It is intended that the book, which is aimed at young and graduate engineers, technicians and those enrolled on higher education courses, will provide sufficient information to enable a designer to understand and overcome the difficulties which will often be encountered during the course of a particular construction project, from the construction of a new leisure complex or housing/industrial estate through to infrastructure design on an altogether larger scale.

Acknowledgements

I am extremely grateful to a number of colleagues who have contributed in different ways during the preparation of this book. In particular, my thanks go to two reviewers, Bill Jeffries of Southern Water, for his invaluable comments on the drainage aspects covered in this book, and Derek Barton of Owen Williams for his review and comments on the chapters covering pavement and highway design.

1

Design Preliminaries

1.1 Introduction

A methodical and analytical approach to the design of any project is an essential element if it is to have a successful outcome.

Without adequate information, little worthwhile detailed design on any project can be progressed and the resulting guesswork will mean that a designer is likely to produce either an under-design or an over-estimate of the requirements on the project. This can result in delays and excess cost to the project and can give rise to unnecessary contractual claims.

Prior to undertaking detailed design, a designer must first become familiar with the scheme and must assess all the available information. For additional background information, the designer should also endeavour to look into the scheme environment and site history.

The designer should then identify any specific constraints on the site along with performance requirements and should formulate an approach to the design bearing in mind the timescales and fees available and the programme for construction.

Careful assembly and preparation of this preliminary information will lead the designer to approach the design in the most economical and practicable manner and this will lay the foundations for the design process and will lead to a successful conclusion of the construction project. The following key stages are discussed in this chapter:

- assessment of information
- preliminary identification of specific requirements
- consideration of specific constraints
- consideration of performance requirements
- approach to design
- additional testing and investigations

- costs, confidence and the pro-active approach

1.2 Assessment of information

Prior to giving consideration to any design works, the designer's first task should be to read through and digest the information presented at the outset.

The purpose of this is to allow the designer to get a feel for the job by disseminating the information and picking out salient and relevant points, making notes where appropriate so that other members of the design team can refer back to them later as necessary.

At this stage, the designer's aim should be to formulate an overview of the scheme and to become familiar with its basic elements, for example, site specific requirements, existing site conditions and constraints, performance requirements and timescales.

A standard pro-forma which gives the designer a list of items to be considered can be helpful, but can lead the designer into overlooking those items which do not appear on the list. An example of a typical pro-forma is shown in ***Box 1***. The example is not intended as an exhaustive list, but is a general guide to those items which the designer should bear in mind during preliminary assessment.

It is important for the designer to be able to establish a train of thought such that the design process can be worked through methodically. Approaches, practices and office procedures will vary throughout the construction industry, but the interpretation and use of the information should be the same whatever the project and wherever its location.

1.3 Preliminary identification of specifics

While making a preliminary assessment, the designer should be aware of potentially 'grey' areas. These are areas where information is lacking or unclear and where further details or investigations are necessary. These areas should be prioritised for clarification later in the design process.

- ☐ *Check site investigation report for completeness*
- ☐ *Consider contamination and remediation*
- ☐ *Identify principal elements of design*
- ☐ *Identify interfaces with other developers and/or third parties*
- ☐ *Consider specific performance requirements*
- ☐ *Consider programming/phasing: design programme and overall scheme programme; timescales for completion, enabling packages*
- ☐ *Check for specific access requirements and/or restrictions, e.g. traffic management, haul routes, restrictions on working hours*
- ☐ *Consider outline construction methodology*
- ☐ *Consider the Construction (Design and Management) Regulations 1994, designer's duties, risk assessments; consider hazards and risks*
- ☐ *Surface water and foul drainage: consider gravity or pumped systems; private or adopted or both*
- ☐ *Highways: private or adopted or both*
- ☐ *Consider implications of restrictions on surface water run-off*
- ☐ *Consider the need for discharge consents, land drainage consents, trade effluent licences*
- ☐ *Check for applicability of any relevant legislation and comply with requirements, e.g. Environment Act, Water Industry Act, Party Wall etc. Act*
- ☐ *Consider the need for liaison with statutory bodies such as the Environment Agency, Local Authority, Water Authority and/or Sewerage Undertaker, Highway Authority, English Heritage*
- ☐ *Identify any restrictions/conditions/agreements imposed by planning and other authorities or those already in place*
- ☐ *Codes of Practice, technical manuals, specifications, and British/European Standards etc. Does the designer possess or have access to the relevant and current publications?*

Box 1 *Example of a typical pro-forma or checklist for designers*

In determining the principal elements, the designer should be looking to identify only the main areas of work and not the minutiae of detailed design. An example of a simple statement which identifies the basic elements of a project is illustrated in ***Box 2***. The statement is sufficient for an early appraisal of a small industrial type development.

At this stage of design, this level of detail is generally sufficient to formulate an approach to the design. In the example, principal elements are identified and an immediate overview of the site is given to the reader.

The site occupies part of the grounds of a former educational establishment and the project is to construct a small industrial complex.

Works comprise the construction of one steel framed shed with a ground footprint of approximately 7500sq.m, 2500sq.m of overlay of existing concrete pavement to provide a new lorry park, 5000sq.m of new car park construction, 300m of access road and associated surface water drainage.

Survey levels indicate that the site falls in a northerly direction.

Box 2 *Identification of the basic elements of a project*

1.4 Consideration of specific constraints

Each project will have its own site specific considerations and the designer will have to make decisions based upon the available information, local knowledge and his or her own experience; or by drawing upon the experience of others in the design team.

Site specific constraints can encompass a diverse range of subjects such as archaeological interest, sites of special scientific interest (SSSIs), the presence of protected plant or animal species, the presence of dissolution features such as swallow holes and cavities or even contamination arising from a previous site usage such as a Victorian hospital burial ground.

The designer needs to identify which constraints will have an effect that may impose restrictions on the methods for construction, the timing of the works and thus the programme and which may therefore have a direct effect on costs.

The designer must also assess the extents to which any additional investigations need to go in order to mitigate these possible effects.

1.5 Consideration of performance requirements

Before the principal elements can be considered in detail, the designer must first identify the performance requirements on each element and consider their implications on the methods available for construction.

The designer should have already identified the principal elements from the supporting information and should know what the performance requirements are on each element.

Specific requirements may already be known, for example from previous dealings with the client but generally, specific performance requirements will be found within the documentation presented at the outset of a project such as the design brief, Employer's requirements or a Performance/General Specification.

As indicated previously, the ideas which a designer is likely to consider during the preliminary stages will cover topics such as materials, construction methods and options, construction costs, fees and timescale for completion. Each of these topics will be further broken down as the design progresses and will be influenced by the performance requirements of each element.

For example, the designer may first consider that although one construction option will provide an adequate solution to a specified design life, it may be either too expensive or too slow to construct; other options may be considered in greater detail before either being accepted as possibilities or rejected as impracticable.

These decisions will be very much influenced by the information available at the time of assessment.

1.6 Approach to design

Following chapters in this book will cover in greater detail the principal issues a designer has to consider before progressing down the route of detailed design.

Each new project will have its own site-specific considerations and although it is true that most of the thought processes behind any design are regularly repeated, it is the different combinations of these thought processes which make each project unique.

If, during the initial appraisal of the project, a requirement for further investigations or information has been identified, then this either should be noted and brought to the attention of the client, for instance if additional fees or costs are likely to be incurred, or should be organised once the exact requirements for the investigations and the information required from them have been accurately established.

Before approaching the design, there must be sufficient information available for the designer to be able to put together a design philosophy. This should indicate to other members of the design team the proposed means of arriving at a solution to address the problems of construction and requirements of performance. It should give an indication of the designer's considerations and thought processes on how the design is likely to unfold.

In the first instance, the philosophy may be a simple statement based upon the information available at the time and it may be qualified by any assumptions the designer has had to make pending the receipt of further information.

Ultimately, the design philosophy should make reference to the principal elements of design, the options available for construction, possible construction methods, and requirements for additional investigations. As the design progresses and more information is received, this can be incorporated so that the philosophy can evolve with the design.

The overall size and content of the design philosophy should be proportional to the size of the project. If the scheme is small, then a short paragraph may be sufficient. More complex projects will require a lengthier description.

The design philosophy should be written clearly, concisely and in simple English. It should not be too technical and should be written so that it can be easily understood by a client or third party who may not have technical background. An example of a design philosophy is shown in ***Box 3***.

> *The site is situated on head deposits overlying chalk which is very close to the surface. The soil report indicates that the ground is liable to frost heave. Depths of construction should take this into account.*
>
> *Areas of existing concrete designated to receive overlay need to provide a durable surface, so ensure that any movement within existing slabs is eliminated prior to overlaying, repair areas of broken concrete and clean out existing joints. The overlay must be of adequate thickness to provide a durable surface for the new lorry park so consider the performance of concrete vs bituminous surfacing and consider how best to reduce or eliminate reflective cracking.*
>
> *Ensure the new car parking area and access roads are of sufficient strength for the vehicles likely to use them. Consider the possibility of stray use by other vehicles, e.g. those using the lorry park; the numbers and types of vehicles for the design life; and the durability of different materials.*
>
> *Survey levels indicate that the site falls in a northerly direction. Points of connection for the main drainage are located to the north-west. Check the feasibility of a gravity sewerage system. There is insufficient soils information to be sure that soakaways will work, but experience of a site located 1km away indicates that it is likely they will and chalk generally has good soakage potential. If soakways are not usable, investigate the possibility of water recycyling instead of on-site storage of surface water.*
>
> *Consider further tests to confirm soakage rates and check the requirements of the authorities with regard to the use of soakaways, viz. Specific requirements for groundwater protection.*
>
> *Consider the time of year construction is likely to take place especially with regard to the placement of any chalk fill and construction of concrete slabs.*

Box 3 *Example of a design philosophy*

1.6.1 Programming

Programming of the works, in terms of both the programme for design and the programme for construction is a further important consideration. The designer needs to take into account the length of time it will take for the design team to prepare and produce drawings and specifications as well as for the construction team to build the project. When programming, the designer should be aware of the man-hours available to carry out the design works within the designer's fees.

In all probability, the designer will be working on several schemes concurrently, and an ability to prioritise and manage effectively each of those schemes is a skill which all designers need to develop.

In the first instance then, the designer needs to identify those elements of design which will take priority and upon which other elements of design will depend.

To achieve this, the scheme should be broken down into key stages and the sequencing of those stages planned so that accurate and relevant information is passed, for example from designer to client or project manager, effectively and on time.

In breaking down the programme for the design works, the designer needs to consider the man-hours required for the key stages and also the time/cost implications, especially as regards fees, of progressing the design if several options for construction have been identified.

An example which highlights the importance of how a design programme should interface with the overall scheme programme is enabling works. These are works which must be carried out prior to the main project but without which construction of the main project could not proceed.

Other considerations of equal importance include sequencing or phasing of the main works, co-ordination with other designers, contractors, services suppliers, utilities/statutory undertakers and local authorities. Lead-in times of certain types of specialist plant and equipment should also be considered, as should lead-in times for materials, the manufacture of particular items necessary for construction and steel fabrication.

During initial programming, designers should be aware that in recent years, some of the utilities, statutory undertakers and local authorities have become notoriously slow in responding to requests for design information, whilst others have managed to keep pace with the needs of the construction industry to respond within reasonable timescales. Applications to such organisations for new supplies and designs or for information relating to existing utilities' apparatus should therefore be made early and the designer should provide as much relevant information to them as possible, for example: details of the intended use of the project upon completion, likely occupancy figures, dates for occupancy and anticipated population.

To assist with programming, there are several software packages available to the designer and these are widely used throughout the industry. The use of software allows amendments to be made to programmes very easily, with critical paths revised, maintained and identified with comparative ease. The designer should bear in mind, however, that the time and effort it takes to produce a computer generated programme should be proportional to the cost of the project. It is not economical, for example, for a designer to spend three days producing a programme to the nearest millisecond when all that is required is to overlay 1500sq.m of concrete slab and to construct three precast concrete ring soakaways.

1.6.2 Interpretation and use of supporting information

The quantity and quality of information presented at the outset of the project will vary greatly depending on the size and complexity of the scheme, but the principal supporting documentation will generally fall into the following main categories:

- design brief
- performance specification and Employer's requirements
- site investigations
- feasibility study and scheme plan
- legal obligations

1.6.2.1 Design brief

In its most simplistic form, the design brief instructs the designer on what he or she has to do. It is also the single most important instruction from the client to the designer and, in an ideal situation, will be written, will form more than one sentence and will describe the intended use of the completed project. It may also indicate certain design parameters.

There have been many cases where a designer has overspent on fees purely from acting outside the original design brief because the brief was either ambiguous or too vague; or the design was progressed upon verbal instruction alone. The designer should therefore act only within the scope of a clear and defined brief.

Unfortunately, the ideal brief is not always received and often the instruction to the designer or the description of the intended use will be a statement such as:

> *'....the external works design shall extend to include the 500 space car park and the perimeter access road, footways and all associated drainage....'*

...and with less experienced developers and contractors, the design brief can be even less informative, for example:

> *'....I've got to put in an access road for these houses... could you do a specification for the car parking as well. Oh, and by the way, I also need the drainage done...'*

In both of the above cases it will be up to the designer to infer from any supporting information and/or design guides and to draw from experience what can be construed as reasonable design parameters and then to put forward appropriate solutions which are likely to match the performance expected by the client, i.e. a reasonable design life, low maintenance and capital (construction) costs, and a scheme which functions as the client intended and which falls within budget.

Prior to arriving at the stage of being able to advise the client of these possible solutions, however, the designer will have considered a range of factors which could have an effect on the overall design. These have already been outlined and are discussed in greater detail later in this and subsequent chapters.

Determination of the specific requirements will point the designer to the relevant design guides, reference books and manuals, but without a specific and clearly defined brief, this can be difficult to achieve.

It is important that the designer communicates any variance, ambiguity or request for clarification of the brief or specification back to its originator, either via a project manager or directly if appropriate.

This request should preferably be in writing as this will help to resolve any future disputes should they arise. If this procedure is followed and evidence of any variations is available in writing, then this will minimise the risk of claims arising and also the risk of disputes within or between the design teams.

This is not to say that a designer should not use his or her own initiative to suggest minor changes in order to achieve a more practicable solution, but the designer should not go as far as to suggest designs or to make recommendations which are clearly outside the brief or specification, or which are a clear deviation from the initial concept of the project.

In an ideal situation, the designer will have all the information necessary to produce a workable design solution; however, if the design has to be carried out without sufficient or accurate information, then this should be brought to the attention of the client prior to the preparation of any designs.

The designer should also advise the client of the possible consequences of proceeding down this route and should be able to inform the client of the modifications to the design which could arise upon receipt of the outstanding information.

If this scenario is taken several stages further such that the design has progressed to its closing stages and the designer finds that the brief cannot be practicably constructed as originally described, then a referral back to either a project manager or client for a decision is the recommended course of action.

This situation, of course, should never be allowed to arise, but if the initial design has been qualified by assumptions derived from the information available at the time of assessment and the client has previously sanctioned this approach, then this situation can arise and with reasonable grounds to request additional fees for any necessary re-design.

Further, this course of action effectively means that the designer is asking the client to bear any element of risk without recourse: it would be better to suggest suspension of any detailed design works until the required information becomes available or to undertake elements of design that are not dependent upon outstanding information.

The designer should also be aware of his or her duties under the Construction (Design and Management) Regulations 1994 under which, designers are individually responsible for their designs and following the previously mentioned course could incur action directly against the designer under these regulations.

Notwithstanding the above, there are occasions where time is of such importance that awaiting further information would significantly delay the scheme; for example when construction needs to be complete prior to the occurrence of known tidal or site/weather conditions.

In this instance, the designer should firstly carry out those designs which will be least affected by any changes brought about by the awaited information and should also consider what effect a last minute change in design is likely to have, not only on the remaining elements but also on scheme progress, costs (to the project as well as to the designer's fees) and on co-ordination with other works.

No unsubstantiated design should ever be incorporated into the permanent works and even if the client has agreed to accept all potential risks and claims, this should be strongly resisted.

1.6.2.2 Performance specification and Employer's requirements

In essence, a performance specification will dictate to the designer the performance criteria for the various elements of design. The Employer's requirements will dictate to the designer specific criteria to which the project should conform.

Understandably, there is an overlap between the two and this can lead to contradictory or conflicting clauses, especially if the Employer's requirements have not been made site specific.

To understand these documents better, a performance specification will lay down specific criteria for say, the design life of a road. It will dictate to the designer how the constructed item should perform in use and will indicate the design parameters that are to be used and will determine the options available for the different elements of design.

The Employer's requirements, on the other hand, often stem from when an employer is engaged in the development of several sites over a period of time, for example a chain of hotels, theme pubs or leisure centres. The

Employer sets down a certain set of rules to which each development should conform with minimal variation, thus ensuring a common theme between each development.

On larger schemes, where there will be a variety of disciplines, the specification and Employer's requirements will usually be broken down into those different disciplines, for example, building services engineer, structural engineer, architect, landscape architect, civil engineer etc.

The designer should not follow blindly the parameters set down in either document – this course of action may, on occasions, lead the designer into an over- or under-design.

Instead, the designer should be aware of design parameters, approved codes of practice and technical guides other than those referred to but which are also relevant to the project and should advise the client and others in the design team accordingly if it would be more appropriate to apply those alternatives.

The designer should bear in mind that it will always be easier to convince the client of the feasibility of any amendments if the suggestions put forward will either save the client money or will have little or no cost implication.

1.6.2.3 Site investigations

The extents of any site investigations included as supporting information will vary greatly from project to project, and may or may not be recent, depending upon how long the scheme has been around.

It is often the case on smaller projects that any site investigations undertaken were the minimum required to establish feasibility; whereas on larger schemes, the level of information tends to be much more in-depth.

On occasions, such information simply does not exist and the designer will have to commission independent investigations or make assumptions based on local knowledge, past experience, geological maps etc., prior to commencing design.

The designer will need to ensure that any site investigation information is adequate and appropriate. In the early design stage for most aspects of external works roads and drainage, the minimum information requirements will be details of CBR values, depth to water table, soil classification and existing sewer information/local authority drainage records.

This information will enable the designer to produce basic construction depths for external works/road design and to assess the options for disposal of surface water and foul drainage.

Any requirement for additional information will depend upon the intended usage of the site, the proposed construction methods and the existing ground conditions.

This is discussed in greater detail later in this chapter.

1.6.2.4 Feasibility study and scheme plan

The feasibility study can encompass a wide variety of topics and will vary in detail and depth of information. Depending upon the extent and nature of the client's brief, a good feasibility study should discuss as a minimum the following:

- surrounding road network
- traffic impact assessments
- environmental impact assessments
- site history and previous land use
- specific site requirements or constraints
- planning conditions and constraints
- drainage and services
- contamination and soil investigations

It may also make recommendations with regard to the intended usage of the site which are qualified by the findings of preliminary investigations already undertaken or assumptions based thereon.

The scheme plan and drawings will often be little more than concept drawings which have received planning permission and which have been prepared by the client or by the client's appointed or in-house design team.

Consequently, it is probable that the project will have received little more than cursory glances at the detailed design and engineering aspects and it will be up to the designer to provide an engineering solution based upon these drawings.

The scheme plan and drawings will form the basis of the final design and should be followed as closely as possible; although as conceptual drawings, minor amendments are generally accepted.

Significant changes to any part of the project should, however, be agreed with the client or project manager and brought to the attention of other members of the design team before any detailed design is undertaken on that element.

As regards acceptable changes, for any given project there is generally a greater degree of flexibility in external works, roads and drainage than there is in, say, the architectural or structural elements.

It is possible that the feasibility study will indicate possible legal obligations. These may not have been considered in detail at the concept stage. Legal obligations are discussed later in this and subsequent chapters.

In any project, there will be specific elements which require designs to minimum standards. These standards may not have been considered in detail at the concept stage or the standards may have since changed.

Highway design standards are particularly relevant if the road is to pass over into local authority ownership upon completion; however, county and local authority standards are often subject to change without notice and consequently may differ from those indicated on the original concept drawings.

Additionally, communication between departments within the local authority itself may not always be evident and an estate road layout which may have been agreed with the planning authority may be rejected by the highway authority due to a lack of interdepartmental consultation.

In schemes where roads are to be maintained privately, there can to a certain degree be a relaxation in some of the design codes, but practicality in use should be considered of prime importance and the designer should pay particular attention to items such as vehicle turning circles, refuse and emergency vehicle access and traffic/pedestrian movement.

Amendments to other areas of external works may arise as a consequence of other changes, for example the re-routing of footpaths. A further commonly amended area of external works is the provision of parking spaces.

For example, if the client requests that a minimum of 400 spaces is provided in a given area and the concept drawings indicate a suggested layout, then it is very likely that there will be several other ways of achieving this which may be more efficient.

Overall, there is no hard and fast rule to determine the level of acceptability of amendments proposed by the designer to the original scheme drawings, but the designer's creativity should not extend to making changes for the sake of making them and nor should they go so far as to require further approval by the planning authority or to be a significant departure from the original scheme drawings.

The designer should therefore only suggest alternatives which are improvements (i.e. those which are more efficient, which identify cost savings or which provide a better engineering solution) and which bring about the minimum change required to achieve practicability in construction and use.

1.6.2.5 Legal obligations

There are many pieces of legislation which the designer will encounter over the course of a career. Many of these are discussed in the following chapter, but at this stage of design there are three Acts of Parliament of which the designer should be aware and which are probably the most common forms of legal obligations that the designer is likely to encounter at the outset of a new project.

These are:

- Town and Country Planning Act 1990
- Highways Act 1980
- Water Industry Act 1991

The above Acts are discussed more fully in Chapter 2 but there are specific provisions arising from them which are relevant to the designer at this stage. Evidence of agreements or obligations under these Acts is likely to be included in the supporting documentation.

One of the commonest forms of agreement arises from section 106 of the Town and Country Planning Act 1990. An s106 agreement allows a local planning authority to secure from a developer a binding contractual obligation which effectively is a price for granting planning permission.

If the developer is not the owner of the land and planning permission will only be granted subject to an s106 agreement, then the owner of the land needs to be persuaded to enter into a contractual obligation to bind the land. Such an obligation is known as a local land charge.

In layman's terms, section 106 allows the planning authority to stipulate that if a development is to be granted planning consent, then the developer must agree do something for the local authority in return and a contractual obligation arising from an s106 agreement may:

- require a sum of money to be deposited with the local planning authority
- require that specific operations are carried out on, in, over or under the land
- require that the land is put to a specified use
- restrict the development or usage of the land

This 'work in return' often takes the form of a junction improvement scheme, construction of a new library wing or some other project of a similar nature, although technically, as section 106 of the Act states, these additional works can be:

> '...*such incidental and consequential provisions (including financial ones) as appear to the local authority to be necessary...*'

It is often a requirement of planning that these works are complete before the project under construction is occupied or put into use. The designer will have to take into account these additional works and incorporate their design into the programme.

As outlined above, a section 106 agreement can request either a consequential or financial provision: by consequential, this means that if the development will generate a significant increase in the amount of traffic at a particular junction, the developer may be required by the local authority to construct a junction improvement scheme; and by financial, if the same example is assumed, the developer will be required to provide funds to the local authority to enable the local authority to construct the junction improvement scheme.

The Highways Act 1980 also plays an important role. This Act gives highway authorities the power to enter into agreements with other parties, for example developers.

There are several sections of this Act which are relevant to designers, but commonly included in the documentation supporting a project will be evidence of an agreement under section 278. This will have arisen from negotiations within the planning and highways departments of the local authority.

Section 278 of the Highways Act 1980 can request a contribution towards highway works from a person deriving a benefit from them. An example of the application of this section of the Highways Act 1980 is shown in ***Box 4***.

The Water Industry Act 1991 has similar provisions to the Highways Act 1980 but applies to the water and/or sewerage undertakers. As before, there are many sections of this Act which are relevant to a designer but most are unlikely to be included in the supporting documentation.

Quite often when a new project is to be constructed, diversion of a public sewer is required and this is carried out under section 185 of this Act. The relevance of s185 of this Act is therefore likely to have been identified at an early, conceptual or feasibility study stage and applies when it becomes necessary to divert a public sewer and a 'building over' agreement cannot be entered into under the terms of the Building Act 1984.

'Building over' is a term applied when a public sewer passes within three metres of a new building. It will not usually be permitted on new developments and exceptions to this rule are generally restricted to small extensions to private dwellings.

To understand this better, for a public sewer to pass through land in private ownership, an easement is required so that the sewerage undertaker maintains a right of access across that easement for the purposes of maintaining the sewer.

The easement is generally three metres wide measured horizontally from the centreline of the sewer but in extreme cases, a relaxation in the width of the easement is permitted. If the new development falls within that easement or on top of the sewer, then diversion of the sewer is required and s185 of the Act comes into play.

No part of a building should be constructed within an easement and this includes foundations and basements, but generally excludes fences and driveways. Boundary walls situated within an easement are generally not acceptable and should be replaced with demountable fences. Pipes and services are permitted to cross the easement but should be perpendicular to it.

Information relating to easements and the positions of services including sewers should be passed on to any new owners of the site once construction has been completed; and this information should be included in any Health and Safety file.

There is legislation currently under consideration which would require the inclusion of such information and information relating to all underground services to be indicated on plans at the Land Registry.

A developer owned a site adjacent to a busy road junction and planned to build a small housing estate which required access onto a major road of that junction.

Independently and irrespective of the developer's plans, the highway authority had planned to improve that junction by constructing a roundabout.

The highway authority was of the opinion that the developer was going to benefit from their proposed junction improvement scheme because it meant that access into the proposed housing estate was going to be improved.

The highway authority therefore sought to enter into an agreement with the developer under section 278 of the Highways Act 1980 to secure a financial contribution towards the cost of the roundabout construction works.

Box 4 *Case study of a section 278 (Highways Act 1980) agreement*

1.7 Additional testing and investigations

The information required for the design of external works and pavements is not hugely scientific; generally, designers rely upon information derived from empirical methods.

Any additional requirements for information which the designer has identified during initial assessment will vary widely from scheme to scheme and will arise principally from the options considered for construction, although constraints imposed by legislation, natural obstructions and surface features, the condition and existing hydraulic profiles of receiving sewers and soil types/classification can also give rise to the need for additional investigations.

When commissioning any additional tests, the designer should be specific about what information is required and should be clear as to how the results will be used and interpreted.

To this end therefore, the designer should have a good knowledge of the methods associated with the proposed construction techniques so that the appropriate information can be sought.

Further, tests should be specified which will provide information that will enable the designer to decide whether or not that particular construction method is suitable; for example requesting that during borehole investigations, SPT values are recorded at set intervals below ground level prior to suggesting to the contractor that it will be possible to drive 6 metre long sheet piles as part of any temporary works.

With external works design, the minimum requirements should include information on CBR values, underlying soil conditions, depth to ground water table, and frost susceptibility.

The construction methods employed in roads and external works do not generally require significant additional information and a satisfactory design can often be obtained from this information alone as long as the designer is aware of the performance requirements of the finished product.

If the site investigations reveal particularly low CBR values or poor soil conditions, then the designer will have to consider what methods will be suitable to improve the sub-grade.

For example, will a simple capping layer suffice, or will other improvement measures be required, for example dynamic compaction, cement/lime stabilisation (the use of hydraulic binders), or surcharging of the sub-grade?

It is possible that any improvement techniques put forward will require information which is not available at the outset and the designer's aim should be to provide a specific brief to the contractor undertaking the site investigation and should furnish the contractor with as much relevant information as possible.

As a minimum, this should include the following:

- intended use of the site
- why the investigations are required
- how the information is going to be used
- a contingency plan if the information received means that the construction options need to be reconsidered

In the last item, the benefit of having a contingency plan is that if the information comes directly from site while the investigations are being carried out, then the designer can instruct further investigations to be carried out or can alter the investigations as they progress should this become necessary.

Foul and surface water drainage design can range from complex hydraulic calculations to the application of simple factors for the approximation of rates of foul flows and surface water run-off.

Often a flow study will be required to determine the allowed rate of discharge of surface water from a new development into an existing public sewer. This is generally undertaken by the local authority/sewerage undertaker at their request and at the developer's expense.

A flow study monitors the existing flows within the receiving sewer over a period of weeks. Naturally, this period needs to incorporate a reasonable period of rainfall.

During the study, data logging equipment is located within the sewer at a strategic point or points downstream of the proposed point of connection. From the results, the local authority/sewerage undertaker will determine the capacity of the sewer and thus indicate the allowed discharge into it.

The designer should be aware that other legislation may apply to the site; alternatively, the site may be adjacent to or may itself be designated a Site of Special Scientific Interest (SSSI), an Area of Outstanding Natural Beauty (AONB), conservation area or similar.

Additional investigations into these areas should be carried out as necessary and should determine how these factors are likely to affect the proposed works, for example the construction programme, construction methodology, costs and legal obligations.

Contamination testing should also be considered by the designer if preliminary site investigations identified the presence of contaminants or if there is a perceived risk of contamination arising from former activities on the site.

Further investigations should provide information on the nature, extents and levels of contamination so that remedial works and precautions can be specified.

1.8 Costs, confidence and the pro-active approach

For a design to reach a successful conclusion in construction, a team effort is required and the designer should adopt a professional approach to the design. To achieve this, the following key points should be noted:

- designers should be pro-active with the client and other members of the design team
- adequate liaison and interaction within the design team is of paramount importance to avoid errors and misunderstandings
- designers should be aware of the cost implications of any construction options and should be knowledgeable about each
- designers should be aware of the fallibility of site investigations
- designers should be aware of the available methods of cost control
- designers should consider other works when programming

Once the design methods have been decided and the preliminary design has commenced, it is then far better for the designer to take a pro-active approach and to discuss with the client a preferred solution or range of solutions than to proceed to detailed design and to present the client with a 'fait accompli'.

The former approach will allow the client to see that the design is progressing in the right direction and will give the client an opportunity to comment on the preliminary suggestions. Further, it will demonstrate to the client competence on the part of the designer if the designer can talk knowledgeably on the pros and cons of the proposed solutions.

A pro-active approach should not extend to leading or to allowing the client to make decisions which are not based upon sound engineering judgement.

The designer has a responsibility to present workable solutions and to be knowledgeable about the features, benefits, costs and pitfalls associated with those solutions. In this way, the designer can guide the client to make a balanced and informed decision.

These steps will build the client's confidence in the design team; indeed, when a client is faced with a new team, it will be these first steps which will form the basis of a stable and professional client-designer relationship and which will ultimately contribute towards successful team completion of the project.

The pro-active approach should also be present between the designers from the different disciplines. This will reduce the scope for errors and misunderstandings, particularly when there are several disciplines involved in the scheme. Further, the scope for error increases when there is an interface between design teams and contractors of different disciplines and if discrepancies between the designs remain undetected until the construction phase, there can be considerable cost implications.

Liaison and interaction between the design teams is therefore of paramount importance if this risk is to be eliminated. It will also result in a smoother interface on site at the start of the construction phases.

On smaller projects, or on schemes where the designer has few variables to consider, the preliminary design stage is likely to be very close to the detailed design stage and the designer will not have the scope, in terms of fees, to consider and explore several options in great detail.

In this instance, the best approach is to put forward a preferred solution but to highlight to the client other feasible options and, if appropriate, to suggest a review of the fee if the client requires that the alternative solutions are to be investigated further.

Higher capital (construction) costs may provide a better quality product which will result in a reduction in long-term maintenance costs over the design life of the completed project, but this may not always be a prime consideration for the client and is therefore not always the best solution.

It may be, for instance, that the client has plans to dispose of the completed project at some point in the future after having forecast and taken maximum profitability over a predetermined period. Conversely, the quality of materials need not be sacrificed if the client is intending long-term usage and the use of higher quality materials does not force the project beyond budget.

The designer will need to strike a balance between quality and costs which will be acceptable to the client. Over recent years, the evolution of the design and build type contract has allowed contractors to undertake construction with less regard to the quality of workmanship and materials than would be expected under a more conventional contract such as Institution of Civil Engineers Conditions of Contract or JCT.

This reduction in quality has been enhanced by the very nature of the design and build contract which does little to discourage cost cutting.

The performance specification and Employer's requirements will generally provide a fair indication as to what level of quality is expected, as will experience of previous similar projects; but as stated previously, the designer should be aware of design criteria other than those referred to in the specification and can use these alternatives to identify cost savings.

If the client is, for example, a housebuilder or a developer who has a track record of similar projects, there may be across-the-board in-house standards which are applied to all projects and therefore it is likely that most of the areas where cost savings can arise will have already been identified. Even if suggested alternatives offer only a minimal saving in one area, the cumulative effect may be quite significant; especially if the savings can follow through other schemes.

Occasionally, an enquiry from a client will be along the lines of *'How much will it cost to do x?'* or *'How much can I do for £x?'*

In this instance, typical questions the designer should be asking are, does the client have a budget figure in mind, or is it necessary for the client to seek funding from other sources?

This type of enquiry is typical if the client is a local authority or an organisation where a referral back to a committee is required to secure funding. Where this is the case, capital costs rather than design life or maintenance costs frequently determine whether or not the scheme will go ahead; usually on the grounds that maintenance is cheap but capital expenditure is not and takes a large portion of an annual budget.

For example, the budget for an education department may be sufficient to allow a school to overlay a deteriorating playground but the design life of the overlay which is determined by its thickness and the materials used in its construction will be determined by the available budget.

Additional costs arising from unforeseen circumstances can be reduced if adequate site investigations are carried out prior to any works commencing on site, but inevitably not all circumstances can be predicted. Even though the site investigation may have included ten boreholes and twenty trial pits, there is no guarantee as to what will be found in the spaces between them. There will often be clauses in the Conditions of Contract which make allowances for claims under these conditions, but the Engineer may choose to omit them. For example, with the Institution of Civil Engineers Conditions of Contract, an Engineer may omit or amend clause 12 so that the Contractor takes on board all risks associated with unforeseen ground conditions.

In all cases, whether a conventional contract, design and build contract, construction management contract or a management contract, the designer's proposals will need to fall within a given budget. The emphasis on cost control therefore plays a significant part. There are various methods of cost control, and the software tools to do so are widely used. Each member of the design team should endeavour to understand the likely costs involved for the elements of design for which they are responsible and should be able to control these costs appropriately should the need for variations arise.

1.9 Chapter summary

This chapter has looked at the preliminary aspects of design; i.e. those aspects which the designer must consider before proceeding to detailed design. It has been based upon the taking, using and giving of information and has been set out as ***Figure 1***, designer's flow chart.

A designer is an integral part of a team and as such should adopt a professional and pro-active approach to the project.

To achieve this goal, a designer should approach all designs in a methodical and analytical manner and should undertake basic groundwork in order to provide solid foundations upon which later designs will be based.

An accurate appraisal of any supporting information is the first key stage in the process. From this, the designer can identify specific requirements and can begin to consider any constraints that are likely to be imposed on the project during either the design stage or the construction stage. These constraints may arise from, for example, statutes of legislation, practicability, or physical constraints within the site itself.

Performance requirements will generally be self-evident and are often stipulated, although with experience, the designer will be able to recognise what are and what are not acceptable or reasonable performance criteria. These requirements will lead on to the designer looking at the various aspects of the design itself, having already identified from the supporting information what the key elements of the design will be.

In order to meet the requirements of performance, the designer will then be able to assess what, if any, additional tests and investigations are likely to be required; and can then advise the client on any potential cost implications (i.e. cost savings or additional expenditure), effects on programme and practicability of the scheme as originally proposed.

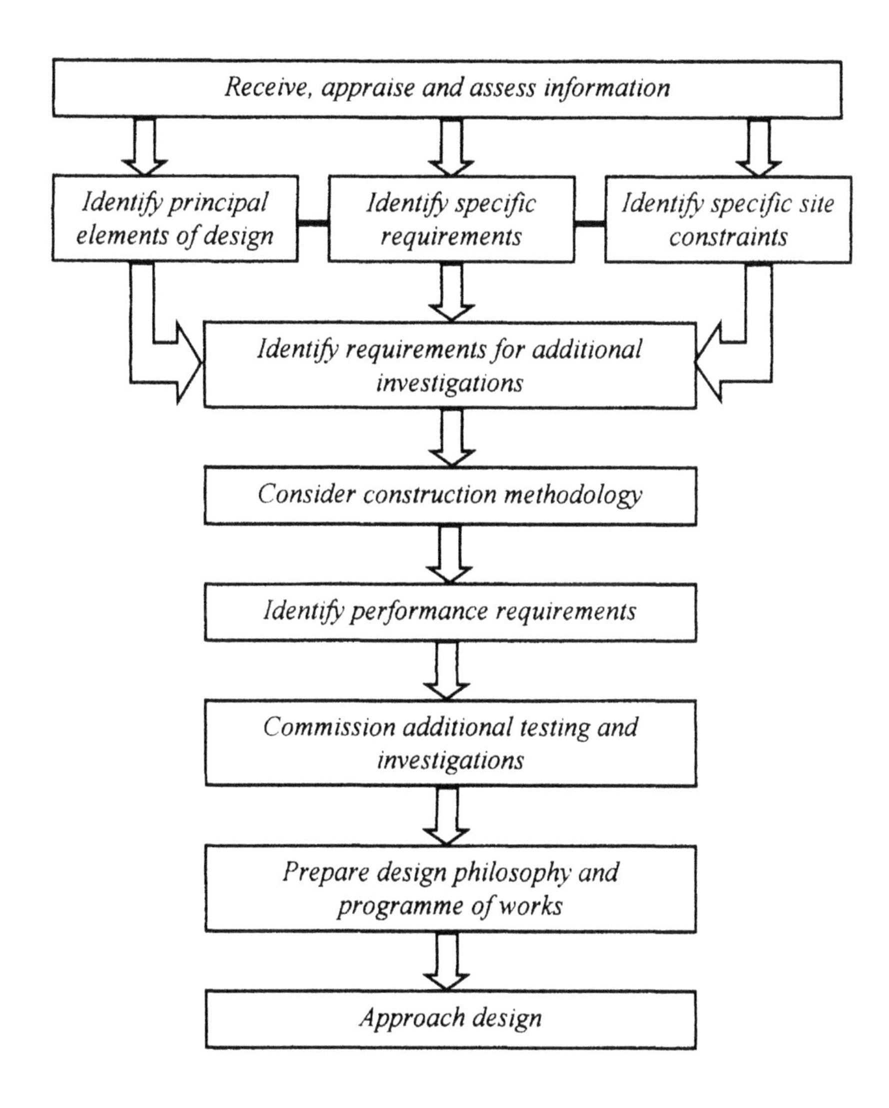

Figure 1 *Designer's flow chart*

2

Essential Legislation

2.1 Introduction

As outlined in the previous chapter, there are many pieces of legislation which the designer will encounter over the course of a career and for any given project, consultations with third parties will reveal what legal obligations and other conditions apply to a site and which have not been identified in the supporting documentation.

Constraints arising from legislation may or may not be indicated at the outset of the design stage - if they are, then the designer will have to assess their accuracy and suitability and may need to argue against their imposition with the relevant body. If they are not, then the designer will have to verify that none are required or are likely to be imposed.

In order to do either of these effectively, the designer needs a reasonable knowledge of the relevant legislation. The most commonly encountered forms of legislation arise from the following statutes. The relevant sections of each Act of Parliament are discussed in subsequent sections:

- Town and Country Planning Act 1990
- Highways Act 1980
- Water Industry Act 1991
- Water Resources Act 1991
- Water Act 1989
- Party Wall etc. Act 1996
- Environment Act 1995
- Environmental Protection Act 1990
- Construction (Design and Management) Regulations 1994

2.2 Town and Country Planning Act 1990

The Town and Country Planning Act 1990 makes provision for the local authority to regulate the development of land. Development can be interpreted as remedial work on contaminated sites as well as the obvious construction of a new building or road. In order for a development to proceed, planning approval must first be obtained from the planning authority. To achieve this, a developer will make an application for planning permission. The application will be determined by the authority in accordance with guidelines set out in a development plan for that authority. There are four possible outcomes of a planning application:

- unconditional permission
- permission subject to conditions
- permission subject to the applicant entering into an agreement under section 106 of the Act
- refusal

One of the commonest forms of agreement arises from section 106 of this Act and was discussed earlier in Chapter 1.

In determination of an application, section 70 of the Act calls upon the planning authority to have regard to the development plan so far as material to the application and to any other material considerations; however, the Planning and Compensation Act 1991 inserts section 54a into the Town and Country Planning Act 1990 and requires the planning authority to determine the application in accordance with the development plan unless material considerations indicate otherwise. These material considerations mean that the issues of contaminated land must be addressed.

In the determination of an application where the site is contaminated, the planning authority frequently requests a section 106 agreement under this Act to undertake remedial works on the contamination to the satisfaction of the planning authority.

2.3 Highways Act 1980

The Highways Act 1980 regulates what can and cannot be done within a public highway and also empowers highway authorities to enter into agreements with others, for example, developers.

The most frequently encountered sections are 38 and 278 and these, respectively, relate to the adoption of highways and contribution to the highway authority towards the cost of certain highway works.

Section 38 comes under Part IV of the Highways Act 1980 and gives authorities power to adopt a highway by agreement. An agreement under this section, commonly known as an adoption agreement, makes a newly constructed highway maintainable at public expense.

For example, a developer wishes to construct a new housing estate but does not want the liability of maintaining the roads within that estate ad infinitum; he therefore offers the roads to the local highway authority after completion and the liability then passes to that authority.

If a road is to be constructed and offered to the highway authority for adoption upon completion, the usual route is via an agreement under this section.

Section 38 states:

> *'... where any person is liable ... to maintain a highway ... a local highway authority ... may agree with that person to undertake the maintenance of that highway... '*

For a developer to enter into a section 38 agreement, the new construction must conform to the highway authority's specification and will generally be inspected by that highway authority (or the highway authority's agent) at various stages, namely, prior to each new layer of road construction from formation level to finished road level. Kerbs, footways and other areas to be offered for adoption will also be inspected at various stages.

The highway authority will also specify the design criteria for sight lines, construction depths, visibility splays, horizontal and vertical alignment, footway and carriageway widths and service margins. These criteria must be adhered to if the road is to be adopted and are usually set down in technical manuals available from the local authority.

Section 278 of the Highways Act 1980 comes under Part XIII of the Act and can request a contribution towards highway works from a person deriving a benefit from them. This section states:

> '...*a highway authority proposing to execute works*...*may enter into an agreement under this section with any other person who would derive a special benefit if those works incorporated particular modifications or features or were executed at a particular time or in a particular manner*... '

An example of the application of this section of the Act was seen in Chapter 1.

Also relevant to the designer are the following parts and sections of the Act:

- Part IV, section 37
- Part V, section 79
- Part VIII, section 119
- Part IX, sections 130, 131, 153, 163, 167, 174 and 184
- Part X, sections 190, 191, 193, 195 and 197
- Part XI
- Part XIV, section 296

Also under Part IV of the Act is section 37. This section provides a means of making a way maintainable by public expense. If, for example, a person such as a developer proposes to dedicate a new road as a public highway, then he can serve a notice on the highway authority to that effect.

The highway authority will then certify that the road has been dedicated as a public highway. For this section to be relevant, there are four conditions which must be satisfied:

- the highway authority must be of the opinion that the proposed highway will offer a sufficient utility to the public to justify its proposed maintenance at public expense
- the highway authority must be satisfied that the proposed highway has been made up (constructed) in a satisfactory manner

- the person proposing to dedicate the way as a public highway must keep it in repair for a period of twelve months from the date of the issue of the certificate by the highway authority
- the way must be used as a highway during that twelve month period

Note that the interpretation of the word 'way' covers, but is not limited to, cycleways, carriageways, footways, towpaths and bridleways.

Part V of the Act relates to the improvement of highways. Within this Part is section 79. This clause empowers a highway authority to prevent the obstruction of view at corners.

This section covers the erection of buildings, fences, trees, hedges and hoardings and is particularly relevant during the creation of new accesses onto existing public highways when it manifests itself under the guise of visibility splays.

There are further references to visibility splays in the *Design Manual for Roads and Bridges, volume 5, Road Geometry*, where design criteria are specified for different categories of road.

The requirements for visibility splays are also set out in design manuals published by highways authorities which are aimed at developers of new estates.

Section 119 falls under Part VII of the Act and applies when there is a public way across a developer's land which it becomes necessary to divert as a result of the intended project.

In this case the developer enters into an agreement with the highway authority and the diversionary works are undertaken at the developer's expense. The developer is also expected to bear the reasonable costs incurred by the highway authority.

Sections 130, 131, 153, 163, 167, 174 and 184 all fall within Part IX of the Act. This Part covers the subject of interference with highways and the powers of the highway authorities to prevent such interference.

Section 130 empowers the highway authority to protect the rights of other users of the highway and to prevent unauthorised stopping up or obstruction of that highway.

This section contains the following catch-all sub-clause (clause 5) which states:

'... a council may... under the foregoing provisions of this section, institute legal proceedings... and generally take such steps as they deem expedient.'

This clause effectively gives the highway authority, within the precedents of law, a free rein to prevent interference with the highway.

The clause can therefore cover excavations and so, if a developer wishes to make a drainage connection, construct a new vehicular access or undertake any opening in the highway, an agreement with the highway authority must first be obtained.

Unfortunately, the Act makes no provision for such an agreement and other acts are called into play by the highway authority for this purpose. One such Act is the Local Government Act 1972 (section 111 empowers a local authority to enter into an agreement which will be to the benefit of that authority). Other authorities may use section 278 of the Highways Act to permit the developer's contractor to undertake the works in the highway.

Section 131 (Clause 1, sub-section (a)) states that the creation of an unauthorised opening (i.e. an excavation) in the highway is an offence.

Section 153 states that no new doors or gates can open onto a public highway.

Section 163 is designed to prevent water flowing onto a public highway from private land. This means that if a new driveway is constructed and it falls towards the highway, any water (e.g. rain) falling on that drive must be intercepted before it reaches the highway. This also applies to new driveways constructed as part of a new development where the road onto which they have access is to be adopted under section 38 of this Act.

Section 167 applies to retaining walls and applies to any length of wall retaining more than 4 feet 6 inches (approximately 1.37 metres) high which is situated within 4 yards (approximately 3.7 metres) of a public highway.

It follows therefore, that any new retaining wall constructed closer than 3.7 metres from the highway and retaining more than 1.37 metres on one side is deemed to be a highway structure and must therefore comply with the requirements of the highway authority in terms of structural design.

Section 174 requires any person executing works within a highway to erect, maintain and remove upon completion any barriers and traffic signs necessary to advise, warn or regulate traffic and to guard and illuminate those barriers and signs during the hours of darkness.

This section is supplementary to the New Roads and Street Works Act 1991 and does not prejudice any of the relevant sections of Chapter 8 of that Act.

Section 184 is the final section in Part IX of the Highways Act which is of particular relevance to the designer. It states that if a development is granted planning permission and it appears that the development will require the construction of a new vehicular crossing over a footway or the improvement of an existing vehicular crossing, then the highway authority can request that those works are executed to their satisfaction.

In theory, the highway authority could serve an enforcement notice under this section but in practice, most developers and contractors are aware of their obligation to construct new vehicular crossings and the costs in proportion to the overall contract value is generally very small.

Also of particular relevance to the designer is Part X which covers new streets. Sections 190, 191, 193, 195 and 197 fall within this Part.

Essentially, this Part sets out that any new streets must conform to certain design standards (e.g. road widths and road geometry) laid down by the highway authority. A council makes byelaws which prescribe these criteria and can vary them to suit the application, but generally, the standards are laid down in design manuals published by highways authorities. There are further references to design standards set out in the Design Manual for Roads and Bridges, volume 5, Road Geometry, where design criteria are specified for different categories of road.

Part XI of the Act covers the making up of new streets and encompasses sections 203 to 237.

Section 205 makes reference to the private street works code and this empowers a highway authority to undertake works in a private street and to apportion the cost of those works between the frontagers of that street. This section is generally enforced when a private street has fallen into a state of disrepair.

To put this in a very simple context, let's say that an existing private street has fifty frontagers. If that street later falls into a poor state of repair, then the highway authority may from time to time make any repairs to that street which it considers necessary. The cost of the repairs is then apportioned between all the frontagers of that street. If there are fifty houses, then the costs are apportioned between those fifty houses.

Subsequent sections up to section 218 cover the conditions attached to the private street works code.

Section 219 makes reference to the advance payments code and this empowers a highway authority to request an advance payment from the owner of land upon which the owner proposes to erect a building which has a frontage onto a private street.

Taking the previously mentioned private street as an example, let's say that all fifty houses and the private street were sold off to a developer and the developer demolished them all and constructed fifty new houses. The highway authority could request a payment from that developer to allow them to make up that private street from time to time in the future should the street later fall into a state of disrepair. To avoid making this payment, the developer has the option of offering the road for adoption under section 38.

Subsequent sections up to section 225 cover conditions which regulate the advance payments code.

Section 296 falls within Part XIV of the Act and is the final section of the Act which has relevance to the designer. This section gives a highway authority the power to execute works on behalf of another person. For example, if there is a section 106 agreement (Town and Country Planning Act 1990) which requires a developer to construct a ghost island at a junction, the developer may enter into an agreement with the highway authority to undertake those works on his behalf. The developer then pays the highway authority the cost of those works just as he would his own contractor.

2.4 Water Industry Act 1991

The Water Industry Act 1991 has similar provisions to the Highways Act 1980 but applies to the Water and/or Sewerage Undertakers and regulates how and what can be connected to a public sewer, and there are several sections which are of particular relevance to the designer. These are sections 98, 101, 102, 104, 106 and 185.

Sections 98 and 101 apply when a sewer is to be requisitioned; that is, a person (often a developer but can be an individual or a group of individuals) requests that the sewerage undertaker provides a sewer so that a site can be drained. Section 101 applies to existing premises.

Section 102 applies when an existing sewer is to be offered to the sewerage undertaker for adoption.

Section 104 arises when new sewers are to be constructed and offered for adoption upon completion and in some respects is similar to a section 38 (Highways Act 1980) agreement, whereby a developer will construct a sewer to a sewerage undertaker's specification with the view to making it maintainable at public expense upon completion.

The current edition of *Sewers for Adoption* gives further guidelines on section 104 agreements and lays down guidelines for the construction of sewers to adoptable standards.

Section 106 applies to the making of new connections to existing public sewers.

Section 185 applies when an existing public sewer is to be diverted.

2.5 Water Resources Act 1991

The Water Resources Act 1991 is a consolidation Act. It specifies what can and cannot discharge into a controlled watercourse and the method in which it can be discharged.

There are provisions in this Act which relate to the earlier Water Act 1989 and these are section 85 (which relates to the pollution of groundwater and surface water) and section 161 which requires action to be taken to clean up the sites causing pollution.

Section 85 makes it a criminal offence to cause or knowingly permit:

- any poisonous, noxious or polluting matter or any solid matter to enter any controlled waters
- any matter other than trade effluent or sewage effluent to enter controlled waters by being discharged from a drain or a sewer in contravention of a relevant prohibition
- any trade effluent or sewage to be discharged
  - into any controlled waters, or
  - from land in England and Wales to a pipe into the sea outside the seaward limits of controlled waters
- any trade effluent or sewage effluent to be discharged, from a building or from any fixed plant onto or into any land or into a lake

or pond which does not discharge directly or indirectly into a river or watercourse (i.e. not a discharge to controlled waters) where the Environment Agency has specifically prohibited such a discharge, or

- any matter whatever to enter any inland waters so as to tend (either directly or indirectly or in combination with other matter which he or another person permits to enter those waters) to impede the proper flow of the waters in a manner leading or likely to lead to a substantial aggravation of

    - pollution due to other causes, or
    - the consequences of such pollution

Water pollution offences are triable either way, that is they can be heard either in a Magistrates' Court or in a Crown Court. The Environment Agency has a discretion to prosecute and the Agency's response to a pollution incident will be governed by its code of practice which currently ranks water pollution incidents into three categories, with Category 1 being the most severe. Category 1 incidents involve at least one of the following:

- potential or actual persistent effect on water quality or aquatic life
- closure of a potable, industrial or agricultural abstraction point
- an extensive fish kill – in excess of fifty deaths
- excessive breaches of consent conditions
- a major effect on the amenity value of the water source, or
- extensive remedial measures are required

The usual consequence of a Category 1 incident is prosecution. Category 2 incidents are less severe, but are still classed as significant and will involve at least one of the following:

- notification to abstractors is required
- a number of fish deaths, not exceeding fifty
- a demonstrable effect on invertebrate life
- the water is rendered unfit for stock

- the bed of the watercourse has become contaminated, or
- the amenity value of the water source has been reduced as a result of odour or appearance

The usual consequence of a Category 2 incident is prosecution or a caution. Category 3 incidents are minor incidents which are unlikely to have significant environmental impact. The usual penalty for a Category 3 incident is a warning letter.

Section 161 of the Act gives the Environment Agency powers to take action or to require action to be taken by others to clear up a potential or existing pollution event. The Environment Agency can also recover the costs incurred in these works from the polluter.

2.6 Water Act 1989

This Act created a single authority, the National Rivers Authority (now the Environment Agency), which is responsible for water quality. Section 107 of this Act made it an offence to cause or knowingly permit the pollution of groundwater and surface water. Section 115 required action to be taken to clean up the sites causing pollution.

This Act was an enabling Act and was superseded by the suite of enabling Acts of 1991 (Water Industry Act and Water Resources Act).

2.7 Party Wall etc. Act 1996

The Party Wall etc. Act 1996 came into force in July 1997 and applies to England and Wales including the Greater London area.

The purpose of the Act is to define the rights between neighbours when one wants to develop land or premises where adjoining the boundary line between lands.

The Act is administered by lay surveyors, known as party wall surveyors and who are appointed to ensure the provisions of the Act are administered with fairness between the parties involved.

There are four definitions of the term 'party wall' as follows:

- a wall where neighbour owners are tenants in common
- a wall which is divided longitudinally into strips, where each neighbour owns one strip
- a wall belonging to one neighbour entirely but where the other neighbour has a right to continue use of it as a division wall between separately occupied properties or properties that can be separately occupied
- a wall which is divided longitudinally into parts with each part being subject to an easement of support for the benefit of the other part

It should be noted that a party wall does not always separate buildings. In case 3 above, it may be a boundary wall. In cases 1, 2 and 4 it may be a party fence wall. Where party walls separate open yards, they are generally referred to as party fence walls.

The most relevant section of this Act is probably section 6 which makes provision for the effects of adjacent excavation and construction. It states:

> *(1) This section applies where -*
>
> *(a) a building owner proposes to excavate, or excavate for and erect a building or structure, within a distance of three metres measured horizontally from any part of the building or structure of the adjoining owner; and*
>
> *(b) any part of the proposed excavation, building or structure will within those three metres extend to a lower level than the level of the bottom of the foundations of the building or structure of the adjoining owner.*
>
> *(2) This section also applies where -*
>
> *(a) a building owner proposes to excavate, or excavate for and erect a building or structure within a distance of six metres measured horizontally from any part of the*

> *building or structure of the adjoining owner; and*
>
> *(b) any part of the proposed excavation, building or structure will within those six metres meet a plane drawn downwards in the direction of the excavation, building or structure of the building owner at an angle of forty-five degrees to the horizontal from the line formed by the intersection of the plane of the level of the bottom of the foundations of the building or structure of the adjoining owner with the plane of the external face of the external wall of the building or structure of the adjoining owner*

Where this section applies, the building owner must serve a notice on the adjoining owner at least one month in advance of the proposed works.

The notice must indicate the proposals and state whether or not underpinning or other strengthening or safeguarding of the foundations of the building or structure of the adjoining owner is included in the proposals.

The notice must be accompanied by plans and sections showing the site, depth of excavations and the proposed building/structure (if applicable).

For the purposes of this Act, excavation can include piling, underpinning and any other works which involve the removal of earth and which will be carried out in any location defined in this Act.

It should be noted that in some circumstances the term building owner may be defined elsewhere e.g. in conditions of contract, as being the contractor.

This is because the permanent works do not become vested in the employer until after their completion. This is a contractual relationship which is relevant to the contract only and the term building owner which is referred to in the Party Wall etc. Act 1996 refers in this instance to the owner of the completed permanent works. This situation is clarified in ***Box 5***.

There are several other sections in the Act which cover the situations where an agreement is required between party wall neighbours, although probably the most relevant to the engineer is section 6 as described above.

> *A sewerage undertaker proposes to construct an underground tank as part of a water quality clean-up operation. The tank is to be sited beneath an open space designated as a public car park and owned by the local authority. The car park is situated adjacent to a privately owned block of flats. The sewerage undertaker proposes to let the contract as a design and build type contract.*
>
> *The contract is awarded and the succesful tenderer commences work to construct the tank. Excavation is 9 metres deep, four metres away from the adjacent block of flats. The contractor asks the sewerage undertaker if the relevant notices under this Act have been served on the owners within the block of flats. The sewerage undertaker's response is that the ownership of the permanent works remains with the contractor until handover and that serving of the notices is therefore the responsibilty of the contractor.*
>
> *Under the Party Wall etc. Act 1996, this is not the case. The sewerage undertaker is the one who wishes to develop the land and the contractor has merely been instructed by the sewerage undertaker to do so. It is therefore the sewerage undertaker's responsibility to serve the notice and to enter into any subsequent agreements with the party wall neighbours.*

Box 5 *Example of the application of the Party Wall etc. Act 1996*

2.8 Environment Act 1995

This Act defines legally the term contaminated land and this, by definition, is:

'...land which appears...to be in such a condition, by reason of substances in, on or under the land that

(a) significant harm is being created or there is a significant possibility of such harm being caused; or
(b) pollution of controlled waters is being, or is likely to be caused'

The Environment Act 1995 created the Environment Agency and transferred to it the powers, duties and assets of the National Rivers Authority (NRA), Her Majesty's Inspectorate of Pollution (HMIP) and the local Waste Regulation Authorities. The Environment Agency is now the principal regulator behind this legislation.

The Act specifies how the quality of land is to be assessed (i.e. the extents of contamination) and the processes by which those parties responsible for remediation costs will be identified. It is a relatively new piece of legislation and land that is likely to be targeted under this legislation will probably include:

- land likely to contaminate crops (this has the potential to cause harm to animals and people eating the produce)
- land which contains significantly hazardous substances at or near surface level (e.g. asbestos fibres and carcinogenic materials and compounds)
- land where substances may affect watercourses and aquifers
- land which emits measurable harmful gases and vapours (e.g. gases and vapours that are flammable, toxic or asphyxiating)
- land subject to subterranean fires

The Act is easier for regulators to use than previous legislation such as the Environment Protection Act 1990 and the Water Resources Act 1991 as they no longer have to risk their own funds in advance of action against the polluter to recover costs.

As a consequence, those responsible for contaminated land or for allowing pollution to occur may find that under this Act, the regulatory bodies will take more interest in their activities and contaminated land (as defined under s78) will be required to undergo remediation.

Judgements will be made by the local authority under the advice of the Environment Agency as to whether or not land is contaminated as defined by this Act.

There are four categories of land which are legally allowable:

- land where significant harm has already been identified
- land where the probability of occurrence of significant harm is high
- land where controlled waters are being polluted
- land where the potential for pollution of controlled waters is probable

The local authority may be responsible for making judgement as to whether or not land is contaminated and for taking the necessary action to ensure remediation is carried out, but there is no obligation for the local authority to bear the cost of the remediation. There are however, three basic principles arising from this Act as regards the recovery or apportionment of costs:

- polluters should pay
- the local authority should not have to bear the costs of those found guilty
- financial hardship should be recognised as a basis for reduced costs

Principally, this Act is written with reference to the polluter and the present land owner, although developers, contractors and their consultants may find themselves exposed to action where development or works increases the risk of contamination, alters the migration pathways of contaminants to cause such a risk or a new risk of occurrence of harm, or causes localised or contained contamination to become spread over a larger area.

2.9 Environmental Protection Act 1990

This Act came about to improve the control of pollution arising from industrial and other processes.

It strengthens powers from earlier legislation for sites causing a threat or nuisance from contamination and sets out the frameworks for IPC (Integrated Pollution Control) and the duty of care with respect to wastes. The aim of both of these (IPC and the duty of care) is to prevent future contamination by using effective control measures.

The Act also covers pollution from former landfill sites by empowering waste regulation authorities to deal with the pollution.

2.10 Construction (Design and Management) Regulations 1994

The Construction (Design and Management) Regulations 1994 came about to improve health and safety for all those involved in a construction project.

Since 1974, clients and designers have had statutory duties to deal with the health and safety issues which affect their works.

Previous construction regulations created the impression that only contractors could affect the safety of their operatives and as a result, clients and designers tended to leave health and safety issues to the contractors.

A study of fatal accidents in the construction industry, however, drew the conclusion that over 60% of accidents were due to decisions made before any works began on site.

The Construction (Design and Management) Regulations 1994 became law in December 1994 and were implemented in March 1995. The equivalent Regulations in Northern Ireland are the Construction (Design and Management) Regulations (Northern Ireland) 1995 (SI 1995/209).

The regulations apply to a wide definition of construction work. They are not restricted to the construction of new buildings or structures or contracts with construction companies and can equally apply interdepartmentally within companies.

The CDM Regulations are based around the general duties of the Health and Safety at Work, etc. Act 1974 and place specific duties upon clients, designers and contractors.

With the coming into force of the regulations, clients, designers and contractors each have to take into account, co-ordinate and manage effectively their respective health and safety issues throughout all stages of a construction project from conception, through to the execution of works on site and subsequent maintenance and even to final demolition and removal. There are four main areas into which the CDM regulations are divided:

- Risk assessment. The parties involved must consider the risks involved and risk assessments, particularly where there is a specific requirement to assess risk. For example, designers cannot comply with their duties to eliminate or reduce risk (regulation 13 of the regulations) unless they have identified and assessed it first.
- Competence and adequate resources. All personnel involved in a construction project must be able to demonstrate suitable competence and have adequate resources to carry out their duties under the regulations.
- Co-operation and co-ordination. The regulations impose a requirement on all those involved in a construction project to co-operate with each other in the interests of health and safety. There is a need to ensure that all those involved co-ordinate their activities to identify and avoid or reduce risk.
- Information. All those involved in a construction project who have a duty to fulfil under the regulations also have a duty to pass on information relating to health and safety.

The following is an interpretation of the definitions in the regulations. Precise and legal definitions are contained within the regulations:

- Person. This means a legal person. It can be a company, a partnership or a group of people. It can also be an individual.
- Client. A person for whom a project is carried out. A client can be someone in-house if he is the one who issues the instruction to carry out the project.
- Developer. The requirements of the regulations which apply to clients

can apply to developers as if they were clients. If a construction project is to build a house for a private client and these works are carried out by another person, then this other person is deemed to be a developer.

- Designer. A designer is a person who prepares a design or who arranges for a person under their control to do so. If the Client makes decisions which affect the design, then the Client can also be counted as a designer.
- Design. A design can include drawings, details, specifications and bills of quantities.
- Contractor. A contractor is a person who carries out or manages construction work or who arranges for a person or persons under their control to carry out or manage construction work.
- Construction work. Construction work includes the carrying out of any building, civil engineering or engineering construction work.
- Structure. A structure is defined as any building, steel or reinforced concrete structure, formwork, scaffolding or other similar temporary structures, or any fixed plant from which a person can fall more than two metres.

The regulations apply to any project involving construction work or any project where five or more persons will be involved at any one time. If the work includes demolition, however, then the regulations will always apply.

Further, the duties of a designer under regulation 13 always apply in relation to construction work. Notwithstanding this regulation, there are several exemptions from the regulations, and these are as follows:

- the regulations do not apply to construction work on premises where health and safety legislation is normally enforced by a local authority (i.e. is not enforced by the Health and Safety Executive). If external contractors are used for construction work, however, the exemption from the regulations only applies if the construction work is of a minor nature
- the regulations do not impose duties on domestic householders, although the duties of any designers involved in the project will still

apply

- the regulations do not apply if the project duration will be 30 days or less and there will be four or fewer persons involved in the work at any one time and the work does not involve demolition. It should be noted, however, that this exemption is in terms of notification to the Health and Safety Executive and the production of formal documentation only. The four items identified on p45 still apply.

3

Consultations and agreements

3.1 Introduction

Depending upon the type of project, consultation will be undertaken by a local authority (or its appointed consultant) or by a developer. In the case of schemes such as traffic calming measures and other local authority projects, the consultations are normally carried out by the local authority involved, although with privatisation and competitive tendering the local authority may have its own consultant to undertake these works on its behalf. In the case of private sector schemes, most of the consultations will be undertaken by either the developer or the developer's consultant.

It would be an impossible task to compile a list of all the consultees for all projects. There are however, several consultees whose names regularly appear and we will concentrate on these for the purposes of this chapter. Some of these are statutory consultees: that is, for example, when an application is made for planning permission then by law the planning authority has to consult specific outside parties, i.e. the statutory consultees. Statutory and non-statutory consultees can include:

- Environment Agency (EA) or Scottish Environmental Protection Agency (SEPA)
- Water and Sewerage Undertakers
- Planning authority
- Highway authority
- Building control
- Utilities and Statutory Undertakers
- Police and emergency services

So why is it necessary to consult these organisations?

Each organisation has its own role to fulfil and comes under its own umbrella of legislation. We have already looked briefly at the main pieces of legislation which the designer is likely to encounter and the following sections will concentrate on the application of that legislation and its relevance to the construction project.

3.2 Environment Agency

In order to fulfil its legislative roles, the Environment Agency needs to be informed of any works which may affect the quality of the environment. Conversely, advice can be sought from the Agency as to how to approach a particular difficulty which may affect the environment.

In carrying out its role, there are thirteen types of plan which require consultation with the Environment Agency, of which only four are not statutory consultations. The plans are:

- Structure Plan – this provides an opportunity for the Environment Agency's concerns and priorities to be reflected as policies and guidance at a strategic level. Statutory consultation.
- Unitary Development Plan – this provides an opportunity for the Environment Agency's concerns and priorities to be reflected as policies and guidance at a strategic and local level and provides the Agency with an opportunity to influence the type, location and scale of development. Statutory consultation.
- Local Plan – this provides an opportunity for the Environment Agency's concerns and priorities to be reflected as policies and guidance at a local level and provides the Agency with an opportunity to influence the type, location and scale of development and provides a vital link to local Environment Agency plans. Statutory consultation.
- Minerals Local Plan – this provides an opportunity for the Environment Agency to influence the location, nature and subsequent operation and aftercare of mineral extraction sites. Statutory consultation.
- Waste Local Plan – this provides an opportunity for the Environment Agency to influence the location, nature and subsequent operation and aftercare of new waste sites. Statutory consultation.
- Air Quality Management Plan – allows the Environment Agency to

assist in the identification and regulation of polluting processes. Statutory consultation.

- Development Briefs – allows the Environment Agency to identify constraints on developments and requirements to protect and enhance the environment. Non-statutory consultation.
- Waste Recycling Plan – allows the Environment Agency to influence the location and management of sites. Statutory consultation.
- Shoreline Management Plan – this is produced by the Environment Agency on behalf of coastal bodies to ensure an agreed and integrated and co-ordinated management of coastal protection and sea defences. Non-statutory consultation.
- Coastal Zone Management Plan – gives the Environment Agency the opportunity to address relevant issues with other bodies and organisations to ensure a co-ordinated and integrated approach to coastal zone management. Non-statutory consultation.
- Local strategies – such as those for the coast, landscape and recreation, countryside, environmental issues, rural areas, conservation and transportation. Non-statutory consultation to influence issues which are relevant to the Environment Agency.
- National Park Management Plan – allows the Environment Agency's interests in management issues to be addressed. Statutory consultation.
- Area of Outstanding Natural Beauty (AONB) Management Plan – allows the Environment Agency's interests in management issues to be addressed. Statutory consultation.

In addition to the above, there are certain types of development which require consultation with the Environment Agency and these, together with the reasons for consultation, are set out below:

1. Developments within or adjacent to any watercourse (including discharges to watercourses). The Environment Agency has a statutory role with respect to all watercourses and consultation is required to ensure that the development does not increase flood flows in watercourses and does not worsen or create flooding downstream and to ensure that adequate access is retained for maintenance. Any works

or activities which involve construction or the modification, erection or re-erection of any structure which may interfere with the bed, banks or flood channel of any watercourse require prior consent from the Environment Agency. Further guidance can be found in Department of Environment Circular 30/92 *'Development and Flood Risk'* and also in the Town and Country Planning (General Development Procedure) Order 1995 Article 10 (1) (p).

2. Developments in areas at risk of flooding. This covers flooding from rivers including tidal lengths and the sea. Areas at risk from flooding are identified on plans provided by the Environment Agency which needs to be consulted to consider and assess the flood risk implications to the existing and proposed development. Further guidance can be found in DoE Circular 30/92 *'Development and Flood Risk'*; PPG20 Coastal Planning; and PPG22 Renewable Energy.

3. Developments which may affect any flood defences. These can include developments on, in, under or adjacent to flood banks, sea defences or any other form of flood defence. Any works within sixteen metres of any tidal defence require prior consent from the Environment Agency. The Environment Agency needs to be consulted to ensure that the stability, continuity and integrity of the flood defence is not prejudiced and that maintenance and emergency access arrangements are maintained. Flood defences are a necessity if serious flooding is to be avoided. Locations of flood defences are available on plans supplied by the Environment Agency. Further guidance can be found in DoE Circular 30/92 *'Development and Flood Risk'*; PPG20 Coastal Planning; and PPG14 Development on Unstable Land.

4. Developments which may affect natural habitats. The Environment Agency needs to be consulted on these types of development to advise on the potential damage to the natural habitat. Any works or activities which will take place within eight metres of the bank of any main river require prior consent from the Environment Agency. The Environment Agency has duties to preserve, maintain and enhance the

environment, flora and fauna and will oppose any development which is likely to have adverse effects on the environment such as the culverting of watercourses. Further guidance can be found in PPG9 Nature Conservation.

5. Developments on contaminated land. On developments such as these, the Environment Agency needs to be consulted so that the risk of pollution of groundwater and surface water and the means of disposal and treatment of the contaminated land can be assessed and appropriate levels of advice provided. Further guidance can be found in PPG23 Planning and Pollution Control.

6. Developments which involve the disposal of sewage other than to a public sewer. This can include sewage disposal by means of cess pits, septic tanks, 'packaged' treatment plant and private sewers and the Environment Agency needs to be consulted to protect surface water and groundwater from pollution. Further guidance can be found in the Town and Country Planning (General Development Procedure) Order 1995 Article 10 (1) (s); PPG23 Planning and Pollution Control; and the Building Act 1984.

7. Developments which may affect groundwater. The Environment Agency has a duty to protect groundwater resources and developments which may have an effect on those resources need to be assessed so that appropriate protection measures can be put forward. The Environment Agency will oppose any developments which create a risk of pollution to groundwater, reduce the availability of the groundwater resource, or impede groundwater flows. Further guidance can be found in PPG23 Planning and Pollution Control; and EC Directive 80/68/EEC – the Protection of Groundwater against Pollution Caused by Dangerous Substances.

8. Developments which are within 250m of a former landfill site. Any developments which take place near land that within the previous 30 years was used for landfill will require consultation with the Environment Agency so that the effects of landfill gas migration can be considered and to assess the potential environmental effects.

Further guidance can be found in the Town and Country Planning (General Development Procedure) Order 1995 Article 10 (1) (x).

9. Any developments which involve the reclamation or raising of land may present flooding or pollution threats. Further guidance can be found in DoE Circular 11/94 The Framework Directive on Waste; and in DoE Circular 30/92 *'Development and Flood Risk'*.

10. Any developments which fall within the Environmental Assessment Regulations 1988. The Environment Agency is a statutory consultee in developments such as these. Further guidance can be found in PPG23 Planning and Pollution Control; Statutory Instrument 1988 No. 1199 Town and Country Planning (Assessment of Environmental Effects) Regulations 1988; DoE Circular 15/88 Assessment of the Environmental Effects Regulations 1988.

11. Vehicle parks. These can include plant hire depots, HGV parks and car parks. Run off from these surfaces can contain high levels of contaminants. The Environment Agency needs to be consulted to ensure that surface water and groundwater sources are not threatened. Further guidance can be found in DoE Circular 30/92 Development and Flood Risk.

12. Petrol filling stations and other chemical storage areas where bulk quantities of chemicals will be stored. The Environment Agency needs to be consulted to assess the pollution threat to air, groundwater and surface water. Further guidance can be found in the Town and Country Planning (General Development Procedure) Order 1995 Article 10 (1) (q); and PPG23 Planning and Pollution Control.

13. Infrastructure schemes. In order to protect the environment from flooding and pollution threats arising from highways projects, railways, airports, tunnels, industrial and chemical processing complexes and other similar schemes, the Environment Agency will assess the proposals. Further guidance can be found in the DoE Circular 30/92 Development and Flood Risk; and PPG23 Planning and Pollution Control.

14. Residential, commercial and industrial developments greater than 0.5 hectares or which include an access road. Developments such as these can present serious downstream flooding threats. The Environment Agency needs to be consulted to assess the implications of the development. Further guidance can be found in DoE Circular 30/92 Development and Flood Risk; PPG12 Development Plans and Regional Planning Guidance; and PPG23 Planning and Pollution Control.

15. Waste management operations such as scrap yards, landfill sites, waste transfer stations and recovery/recycling plants. Air quality, land drainage, groundwater and surface water pollution are at high risk of pollution from these types of development. Further guidance can be found in DoE Circular 30/92 Development and Flood Risk; PPG23 Planning and Pollution Control; EC Directive 80/68/EEC – the Protection of Groundwater caused by Dangerous Substances; and the Town and Country Planning (General Development Procedure) Order 1995 Article 10 (1) (r).

16. Any development within 500m, measured from the site boundary, of any process subject to Integrated Pollution Control or subject to the Control of Industrial Air Pollution (Registration of Works) Regulations 1989. This includes developments within any site where such an activity occurs and consultation is necessary to ensure that pollution issues have been adequately considered. Further guidance can be found in PPG23 Planning and Pollution Control.

17. Developments which could exacerbate existing flooding, sewerage or sewage disposal difficulties. Such developments must not cause additional problems or worsen existing ones and additional infrastructure should be included to ensure this is prevented. Further guidance can be found in PPG23 Planning & Pollution Control; and PPG12 Development Plans & Regional Planning Guidance.

18. Forestry activities may affect water quality and have an adverse effect on existing flood plains. Such activities also have a significant environmental impact. Further guidance is available in DoE Circular

29/92 Indicative Forestry Strategies.

19. Ponds, lakes and reservoirs. The Environment Agency will need to assess the implications of flood risk and environmental impact. This can also include storage of water for irrigation purposes. Further guidance is available in DoE Circular 30/92 Development and Flood risk.

20. Fish farming. Consideration needs to be given to the aquatic environment. Further guidance can be found in the Town and Country Planning (General Development Procedure) Order 1995 Article 10 (1) (y).

21. Cemeteries and crematoria present a significant pollution threat to groundwater and surface water. Further guidance can be found in the Town and Country Planning (General Development Procedure) Order 1995 Article 10 (1) (t).

22. Swimming pools have a potential to cause pollution to groundwater and surface water as the water used in swimming pools contains a high level of chemicals. Any discharge from a swimming pool will normally require a trade effluent agreement and be discharged to a public sewer.

23. Many camping and caravan sites are at risk from flooding and can cause pollution to groundwater arising from foul and surface water discharge and from on-site activities. Further guidance is available in DoE Circular 30/92 Development and Flood Risk; and DoE Circular 14/89 Caravan Sites and Control of Development Act 1960.

24. Developments which may affect access to water or waterside areas. This type of development may have significant impact on the water environment, particularly if there are water based recreational facilities or amenities. Further guidance is available in PPG17 Sport and Recreation.

25. Agricultural and similar developments. Pollution from farm waste has

devastating and far reaching effects. Restrictions may be required on the transport of such waste. Restrictions may be required on agricultural developments and activities such as the grazing of calves and lambs, the storage of chemicals and fertilisers, manure and effluent, intensive farming and silage making. Similar restrictions may extend to stables, zoos, kennels and other places where numbers of animals are kept or sheltered. Further guidance is available in the Town and Country Planning (General Development Procedure) Order 1995 Article 10 (1) (s); and the Control of Pollution (Silage, Slurry and Agricultural Fuel Oil) Regulations 1991.

26. Golf courses may require specific land drainage and water requirements (esp. irrigation). Existing vegetation may be removed, ground profiles altered and high volumes of water may be required for maintenance of the course and this is often achieved by abstraction. Further guidance is available in DoE Circular 11/94 The Framework Directive on Waste; and DoE Circular 30/92 Development and Flood Risk.

27. Industrial developments and process plants which use high levels of preservatives and other chemicals which present significant threats of pollution to groundwater and surface water. Further guidance is available in PPG23 Planning and Pollution Control.

Where a discharge will be required to controlled waters, or where abstraction from controlled waters is required, it is generally advisable to make an application for consent well in advance of the commencement of works due to the time delays involved, principally as a result of statutory notice and consultation periods.

A discharge consent will generally be required for each discharge into controlled waters and the applications should be made to the Environment Agency.

The applications must be published in a local paper and the *London Gazette* by the Environment Agency who must also directly notify the relevant local authorities and water undertakers.

Any written representations must be considered and must have been taken into account within six weeks of the notice being advertised and if the Environment Agency intends to grant consent to the application, it must inform anyone who made a written representation of the decision and then wait a further 21 days.

The Environment Agency has the power to grant consent either conditionally or unconditionally or can refuse the application outright. Generally any conditions will relate to the discharge as follows:

- quality
- quantity
- temperature
- nature and composition
- siting and outlet design
- specific sampling and monitoring requirements

If no decision has been given within four months of the application being made, then the application is deemed to have been refused. In the event of refusal, or if the applicant considers that the conditions imposed on the consent are unreasonable, then the applicant has a right of appeal to the Secretary of State, although a better course of action may be to enter into negotiations with the Environment Agency to reach a compromise prior to approaching the Secretary of State.

Another piece of legislation, the Environmental Protection Act 1990, introduced the system of integrated pollution control (IPC) and this was aimed at improving the regulation of industrial processes. It is also regulated by the Environment Agency and the Environmental Protection (Prescribed Processes and Substances) Regulations 1991 identify six categories of processes, although most are not applicable to typical construction and these, therefore, generally fall outside the system of integrated pollution control.

It can be seen that the Environment Agency plays a significant role in construction and requires consultation on a wide variety of projects, and further to the consultations outlined above, other organisations such as those involved in conservation or archaeology may liaise with the Environment Agency even when there is no statutory requirement.

3.3 Water and Sewerage Undertakers

Water and Sewerage Undertakers have three principal functions to perform:

- the maintenance of sewerage systems and sewerage treatment works
- the regulation of trade effluent discharges
- the provision of drinking water

In addition to these, Sewerage Undertakers have several other duties to fulfil. Firstly, they have a duty to provide an effective means of removing sewage, whether it is foul sewage, trade effluent or surface water run-off. Secondly, they have a duty to maintain and safeguard the means of removing this sewage and finally, they need to give consideration to the future of the sewerage systems for which they are responsible.

Most Sewerage Undertakers will want to encourage orderly development and will actively seek to safeguard their existing and future customers by stipulating what can be discharged to which sewer and in what quantity.

Prior to the introduction of the Water Act 1989, it was a legal requirement for planning authorities to pass planning applications to Sewerage Undertakers as they were statutory consultees. With the introduction of the Act, this requirement ended, although many of the Water and Sewerage Undertakers continue to examine applications and maintain close links with the Environment Agency (who are still statutory consultees as land drainage and sewerage problems are often closely related). This Act was later replaced by the suite of 1991 Acts discussed in Chapter 2.

The role Sewerage Undertakers currently play in planning applications is less than it has been in the past, but in some instances, they may seek to have conditions imposed on a planning consent, or in extreme cases to have the planning application refused.

When designing an outfall from a new development for example, there may be a number of drainage difficulties to be considered:

- foul and surface water will need to be separated
- there may be no suitable public sewers near the site, in which case

the Sewerage Undertaker can be requested to provide one (refer to the later section on sewer requisitions)
- flows from the development may overload the receiving sewer or may exacerbate existing flooding problems

The Sewerage Undertaker will advise on the drainage difficulties likely to be encountered on a particular development and the engineer will need to specify solutions to overcome these.

When an outfall has been decided, in the case of a connection to an existing public sewer, it will be necessary to make an application to the Sewerage Undertaker for permission to make the connection. Under section 106 of the Water Industry Act 1991, a Sewerage Undertaker can refuse to allow a connection if it is thought that the connection may prejudice the operation of the existing sewer.

There are five principal items of information which the undertaker needs at the time of making such an application:

- location of the proposed connection
- whether the connection will be foul or surface water
- the size of the connection
- the address of the properties served by the connection
- the local authority responsible for Building Control

Connections to the public sewer can be made by the Undertaker or by a developer's appointed contractor, but legislation allows for the Undertaker to insist on making the connection itself with the expense being borne by the developer. This power is generally used in circumstances where the connection would be particularly difficult, for example to a rising main, or where the Undertaker has reservations as to the competency of the chosen contractor.

If the connection is in the public highway, permission is not automatically received to open the highway in order to make the connection and is outside the jurisdiction of the Sewerage Undertaker. Under the Highways Act 1980, it is an offence to interfere with the public highway and permission to open the highway to lay a pipe requires permission from the relevant highway authority prior to the connection being made.

Local drainage authorities often hold valuable information which may not be available from the Sewerage Undertaker and may include records of local flooding incidents, rodent infestation, etc. Many drainage authorities have personnel who have looked after the same local sewer and drainage networks for thirty years or more and understand the workings of the system far better than the network administrators in the head office of the Sewerage Undertaker who holds overall responsibility for the area.

3.3.1 Adoption of sewers

A sewer can be built by a developer and offered for adoption and this can happen in two ways. Both methods rely on the Water Industry Act 1991. It should be noted however, that for the sewer to be adopted by the Sewerage Undertaker, the developer, who has a duty to satisfactorily undertake his part of the agreement, must do so.

The first method is to offer an existing sewer to a Sewerage Undertaker without any advance agreement. The way to achieve this is via section 102 of the Act.

There are six principal items of information which must be provided to the Sewerage Undertaker if this is route is to be progressed:

- design parameters for the sewer
- design calculations
- as built drawings which may include a CCTV survey
- details of all connections to the sewer
- land transfer plans, e.g. easements required
- specification for the materials and workmanship used in construction of the sewer

If the information supporting the application is acceptable and if after inspection the sewer is deemed to be of a suitable standard of construction, there are no legal objections and all land transfer plans have been concluded, then the Sewerage Undertaker may adopt the sewer by making a declaration of vesting, although there is no certainty that a sewer offered under section 102 will be adopted.

If for any reason the sewers are deemed to be defective, then the developer has the option of bringing them up to standard, after which a further inspection will be made prior to adoption.

The second route for the adoption of a sewer is by prior agreement with the Sewerage Undertaker. In this case a sewer is constructed to the adopting authority's standard and in compliance with specific guidelines. Currently, these guidelines are laid down in *Sewers for Adoption – 4th edition* and may be amended or supplemented by the undertaker's own safety policy. Details are supplied to the Sewerage Undertaker prior to the adoption and formal approval is given. Works to construct the sewer frequently commence prior to approvals being received and when this is the case, all works are carried out at the developer's own risk.

This second route for adoption brings into play section 104 of the Water Industry Act 1991 and is known as an adoption agreement and the procedure is instigated by an initial communication to the Sewerage Undertaker who then agrees (or not) to adopt the sewer in principle, subject to the necessary approvals.

In the first instance, the developer must undertake to do three things:

- agree to pay all costs reasonably incurred by the Sewerage Undertaker in preparing the agreement to adopt
- supply drawings which give details of the sewers to be adopted
- supply details of all the design parameters and calculations used for the hydraulic and structural performance of the sewer to be adopted

This information is then assessed by the Sewerage Undertaker who then either prepares an estimate of the cost of the works or agrees that the developer's own estimate of costs is reasonable.

The next step is for the developer to provide a bond to the Sewerage Undertaker amounting to a percentage of the estimated cost of the works and there is often a minimum value of bond.

Once this has been paid, the drawings and calculations are then checked and a decision is given on the proposals. If the adopting authority is satisfied, then the agreement is completed and construction can commence;

otherwise amendments to the scheme are requested.

Before construction commences, inspection fees are due to the Sewerage Undertaker and also the fees for preparing the adoption agreement. The undertaker's representative will periodically inspect the works and when completed, the developer should notify the adopting authority by submitting a notice of completion or similar.

The sewer will then undergo a final inspection by the Sewerage Undertaker which will probably include a CCTV survey and if all is found to be satisfactory, then the developer is issued with a provisional certificate.

The sewer then remains the responsibility of the developer for the following twelve months (known as the maintenance period), during which time the developer must maintain and pay for any works required on the sewer. Once the maintenance period has expired, there is a final inspection by the adopting authority and a final certificate is issued if all is found to be satisfactory. The adoption procedure is completed by the Sewerage Undertaker making a declaration of vesting.

3.3.2 Sewer requisitions

Where no public sewers are available, a developer can request that the Sewerage Undertaker provides one. This is carried out under either section 98 or section 101 of the Water Industry Act 1991 and is known as a requisitioned sewer. A person or developer applying for a requisitioned sewer has to agree to fund each stage of the requisitioning process. Section 101 is used for existing properties which are causing pollution problems.

Requisitions are not required if the route of the sewer is in the public domain (e.g. public highway), as the developer can lay the connection with an adoption agreement as he can over private land if he can agree an easement. The purpose of a sewer requisition is to prevent one developer or landowner from holding another to ransom. A connection from one property only cannot be requistioned, but if the connection is to serve two or more properties, then the connection is, by legal definition, a sewer and can be requisitioned. It should be noted that the Department of Transport and Environment for the Regions is currently looking at amending the law so that responsibility for a lateral connection rests with the Sewerage

Undertaker rather than the property owner as it does at present. Such a change may mean that individual connections become requisitionable.

There are several rules which govern sewer requisitions:

- requisitions for sewers are normally accepted only from the boundary of a site and not within the site itself
- rising mains may be requisitioned
- requisitions for pumping stations within a site are not normally accepted
- connections to a sewer cannot be requisitioned
- the route, discharge point and points of connection will normally be determined by the Sewerage Undertaker
- highway drains cannot be requisitioned
- drains carrying only trade effluent cannot be requisitioned

Requisitions for sewers which carry non-domestic effluent will only be accepted if those sewers carry an element of domestic effluent and the developer agrees to accept the cost of providing for the non-domestic portion of the flow.

To make an application for a requisitioned sewer, a formal notice has to be issued to the Sewerage Undertaker. Prior to this, however, it is advisable to have discussed the intention to make such an application with the Sewerage Undertaker as once the notice has been issued, fees become due.

It is highly unlikely that applications for requisitions will be approved where the land in which the requisitioned sewer is to be laid is not in the public domain (e.g. under the public highway) or where the point of discharge is a watercourse and no consents to discharge have been agreed.

There are several items of information which the Sewerage Undertaker needs to know before progress can be made on the application.

- site plan
- the anticipated water consumption for any non-domestic portion of the development (or flow figures for the proposed discharge)
- drainage arrangements within the site
- details of the buildings (residential and non-residential) that are to be

erected on the site
- programme of works and occupation of the building

Once the Sewerage Undertaker has this information, preliminary designs and cost estimates are carried out and a programme of works for the requisitioned sewer is drawn up. Timescale for completion of the works by the Sewerage Undertaker is regulated by the Water Industry Act and is normally within six months of entering into a contract to provide the requisitioned sewer and the agreement of the start and end points of the sewer.

There are generally two methods of payment for a requisitioned sewer. The first method of payment is covered in sections 99 and 100 of the Water Industry Act 1991 and the developer must agree to pay or must provide a bond or surety for:

Either :
- the cost of the entire sewer requisition
- interest on any loans needed to finance the project

Or:
- a fee which covers the relevant deficit for each of the twelve years after completion of the requisition. The relevant deficit is calculated often using the developer's programme of occupation rates on the completed development to calculate the income. The twelve year period is capitalised into a single lump sum payment prior to the start of the work on the requisition. The relevant deficit includes for the capital cost and any loan interest

The second method of payment is three-stage as follows:

- an initial payment
- a deposit for a percentage of the cost of the works
- a final balance

The first payment becomes due after the notice has been issued on the Sewerage Undertaker and covers the costs of preparing the preliminary designs and cost estimates.

The second payment covers the costs incurred by the Sewerage Undertaker in taking the scheme to beyond the tender stage to appointment of contractor.

The final payment covers the costs of the works and the additional fees such as legal costs, compensation costs, design fees, site investigations and services diversions.

3.3.3 Sewer diversions

When a development is to take place on a site where there is an existing sewer and that sewer falls under the footprint of a building for example or no building over agreement can be obtained, then diversion of the sewer becomes a necessity. Section 185 of the Water Industry Act relates to the diversion of sewers.

A Sewerage Undertaker does not have to divert sewers which are in public places such as the highway, but must agree to requests to divert sewers in private land so long as the request is reasonable. Not all requests are reasonable and the diversion of large sewers of strategic importance generally presents the greatest difficulties and also tends to be the least cost-effective for the developer or person requiring the diversion.

There are two ways a sewer can be diverted. The first is where the person requiring the diversion arranges for the works to be carried out. The second is where the Sewerage Undertaker carries out the works on behalf of the person requiring the diversion.

With the former, the person requiring the diversion must indemnify the Sewerage Undertaker for the full estimated cost of the works and it is normal to do this by providing the Sewerage Undertaker with a bond for the full amount. This protects the Sewerage Undertaker should the persons carrying out the works become insolvent.

When preparing a scheme which requires the diversion of a sewer and the person requiring the diversion is arranging for the works to be carried out, there are several principal details which must be provided to the

Sewerage Undertaker. Once these details have been assessed as satisfactory, the agreement can be completed. The details required are:

- drawings of the site showing details of the existing sewer, its proposed route once diverted; a longitudinal section along the centreline of the proposed sewer; construction details of the proposed sewer
- calculations for the proposed sewer based upon design parameters set down by the Sewerage Undertaker
- construction programme and method statements
- an agreement to pay the costs reasonably incurred in preparing the agreement to divert and site inspection and supervision during the diversionary works

Once these details have been supplied to the Sewerage Undertaker, an estimate of the cost of the works is prepared. The Sewerage Undertaker may prepare a cost estimate or may accept an estimate put forward by the person requiring the diversion if it is reasonable.

The calculations and details are checked and a decision is given on the proposal.

If an agreement that the diversion can take place is reached, then the contractor can commence works. Upon completion of the diversion, a notice of completion should be sent to the Sewerage Undertaker who will then make an inspection. If this is satisfactory, then a provisional certificate may be issued by the Sewerage Undertaker.

The maintenance of the diverted sewer remains the responsibility of the person who required the diversion for the following twelve months after issue of the provisional certificate.

Under sections 104 and 116 of the Water Industry Act 1991, the Sewerage Undertaker will ban use of the redundant sewer and make a declaration of vesting for the new sewer, but flows cannot be diverted into the new sewer until written authorisation has been received from the Sewerage Undertaker. A combined agreement under these sections is not strictly legal , as it involves the diversion of existing flows through a new sewer prior to its adoption, thus there is a period when the public sewer run is interrupted. There are obvious benefits to using this procedure however, and the undertaker is protected by the 100% bond, so the risk is often taken.

At the end of the twelve month period, a final inspection is carried out by the Sewerage Undertaker and if satisfactory, then a final certificate is issued and full responsibility for the diverted sewer is borne by the Sewerage Undertaker.

When it will be the Sewerage Undertaker who carries out the diversionary works, in the first instance the Sewerage Undertaker will require the following:

- agreement to pay all reasonable costs incurred in diverting the sewer
- drawings showing the site and sewer to be diverted
- a programme of works for the site

Once the above details have been received and assessed by the Sewerage Undertaker, then an estimate will be prepared and an initial payment will be required from the person requiring the diversion to cover the costs incurred in preparing the preliminary designs and estimate.

A deposit is often required at this stage and usually amounts to approximately 10% of the estimate value and allows the scheme to be progressed to appointment of a contractor.

A final balance will be required to enable the Sewerage Undertaker to complete and commission the works. This amount will depend upon the value of the accepted tender plus costs incurred by the Sewerage Undertaker in respect of items such as survey fees, site investigations, legal costs, services diversions, site supervision and administration of the contract.

3.3.4 Trade effluent consents

Trade effluent can only be discharged into a public sewer with the consent of the relevant Sewerage Undertaker and legislation establishes precise controls for trade effluent consents and agreements. As stated earlier in this chapter, The Water Industry Act 1991, sections 118 and 129, covers this type of discharge.

Trade effluent can be highly toxic and harmful to the sewerage system and the Sewerage Undertaker has a duty to protect the system and the health and safety of those who maintain and are responsible for it. Ultimately, the trade effluent will end up after treatment in a watercourse or in the sea and the effluent must be capable of being treated without detriment to the treatment works.

There are many processes which require consent to discharge trade effluent and treatment of such effluent is a highly specialised field. Some trade effluents may require treatment prior to their entry into the public sewer.

When considering a consent to discharge trade effluent, the Sewerage Undertaker will require three items of information:

- the maximum daily quantity (volume) of trade effluent to be discharged from the process or site
- the peak flow rate of trade effluent from the process or site
- the nature and composition of the trade effluent

If a consent to discharge is received from the Sewerage Undertaker, it is likely that there will be a number of conditions imposed as part of the consent and charges will be levied for the disposal and treatment of the trade effluent. Examples of such conditions can include:

- the stipulation of permitted times when discharge can occur
- the permitted rate of flow of effluent into the sewer
- the nature, temperature and pH of the discharge
- specific monitoring requirements
- record keeping
- the provision, testing and maintenance of specific sampling equipment

The Sewerage Undertaker has power to vary any consent, provided that a two month period of notice is given to the discharger, although this not permitted within two years of the initial consent being granted. Rights of appeal against the variations, conditions imposed when a consent has been granted or against refusal of a consent can be made to the Director General of OFWAT, but there is no right of appeal against the charges levied by the Sewerage Undertakers for the treatment of the trade effluent.

These charges are calculated on the basis of volume and nature (strength) of the discharge.

3.4 Planning authority

As we have already seen, planning authorities can impose restrictions and conditions on any planning application granted permission.

Such conditions can relate to methods and hours of working, phasing of the works, the specification of materials, for example road surfacing materials and colours and can stipulate that works cannot commence until a specific condition has been met, for example the drainage strategy must be approved by the authority.

These conditions will be imposed under the local authority's planning policy and guidelines and will be in addition to any other agreement such as an s106 agreement (Town and Country Planning Act 1990).

In the first instance for a new development or proposed construction activity (this can include a new housing estate, a new road or a simple alteration to an existing building), an application must be made to the local planning authority to request planning permission. This is to ensure that the proposed development falls within the scope of the local authority's own development plan.

A planning authority is obliged to determine an application within 8 weeks of receipt of the application, although where the application is accompanied by an Environmental Impact Assessment, this period is extended to 16 weeks or an otherwise agreed longer period.

Failure to make a decision within the prescribed period may be deemed to be a refusal in which case the applicant has a right of appeal to the Secretary of State. Similarly, if any conditions are imposed which the applicant feels are unreasonable or if the planning application is refused, then again, the applicant has a right of appeal to the Secretary of State.

Appeals to the Secretary of State may take the form of a public enquiry, an informal hearing or a written representation. Most appeals are dealt with by a written representation but larger and high profile schemes tend to go to public enquiry.

The appeals are usually delegated by the Secretary of State to an inspector, but the inspector's report goes before the Secretary of State for the final decision.

If any development contravenes conditions imposed by the planning authority as part of a consent, or if any development takes place without consent, then the planning authority can take legal action, instigated in the first instance by serving an enforcement notice under the Town and Country Planning Act 1990.

This action must be brought about within specified time limits and these are set down in the Town and Country Planning Act 1990 but can be broadly described as 10 years in most cases but 4 years in the case of operational development or changes of use to a dwelling house.

Once the notice has been served by the planning authority, the person on whom the notice was served has two courses of action:

- comply with the notice
- appeal against the notice

Compliance with the notice means that the breach of planning control must be remedied. In the case of a breach of a condition of a planning consent, the person upon whom the notice was served has 28 days to comply. There is no route of appeal against a breach of a planning condition. Appeals against the notice are made to the Secretary of State and operations may continue until the appeal is determined unless a stop notice is served (again under the Town and Country Planning Act 1990) by the planning authority. It should be noted that if there is an intention to commit a breach on the part of the applicant or if there is an actual breach, then the planning authority can apply to the courts for an injunction.

3.5 Highway authority

Highway authorities will have valuable local knowledge of traffic conditions and patterns of movement and should be consulted well before any traffic management proposals are to be put into effect.

The highway authority will be able to advise on any necessary restrictions, particularly with regard to traffic sensitivity, periods of peak traffic flow, carriageway construction and local highway congestion.

Where schemes are likely to require significant traffic management, then consultation with the police and emergency services should also be considered.

It may not always be clear who the highway authority is. The highway authority is usually the county council, although the local authority may be an agent for the highway authority. In a unitary authority, however, the local authority may be the highway authority; and to further complicate issues, in a metropolitan borough, the borough may be the highway authority. The highway authority is responsible for maintaining and overseeing the construction of new roads so that they conform to the county standards.

Liaison with the highway authority will confirm if a proposed scheme complies with these standards, but should have already been undertaken to some degree at the planning stage by the local planning authority.

The highway authority will also be able to advise on the types and sources of features, equipment and materials that are used throughout their area, such as bollards, street lighting columns and other street furniture, surfacing materials and the manufacturers of specialist equipment and materials specified by the county standards or which may be specified as a condition of the planning permission.

3.6 Building control

The building control department of a local authority is responsible for ensuring compliance with the Building Act 1984. This Act basically stipulates how a building should be built. Building control departments often have local knowledge and can be a valuable source of preliminary information, such as local ground conditions, environmental difficulties, drainage problems and they can often give advice on how, locally, these problems have been overcome.

3.7 Utilities and statutory undertakers

The utility companies and statutory undertakers provide services such as sewerage, gas, electricity, water, cable TV and communications and are generally referred to as 'stats' companies. There are two main reasons why utilities and 'stats' should be consulted:

- the proposed works may affect their apparatus
- the proposed works may need new supplies from those companies

When any works are being considered, it will be necessary to consult these companies to ensure that their apparatus is unaffected by the works. Where works are likely to affect such apparatus, then either protection or diversion of the service will be required. In the case of communications, particularly underground fibre-optic cables near hospitals, airports and technology centres, the costs of diversion can be extremely high.

The records provided by the stats companies should be used for guidance only and the exact location of any underground apparatus should be established on site prior to the commencement of any works, firstly for the safety of the operatives working in the vicinity of the apparatus and secondly to minimise the risk of disruption to the service if the apparatus is damaged.

When diversionary works are required the costs are generally borne by the persons requiring the diversion and a similar process to that required for the diversion of a sewer is followed:

- details of the location and nature of the proposed works are supplied to the utility company
- the utility company responds by confirming whether or not diversion/protection works are required
- an estimate of the cost of the works is prepared, usually by the utility company
- in the case of a diversion, a programme is agreed
- in the case of simple protection works, method statements are required from the contractor or are stipulated by the utility company

and a programme is agreed
- the diversion/protection works are carried out
- the development can proceed

Quite often, a new development will require the provision of new supplies so that the users of the developed site have fully serviced units, whether they are industrial units or residential units. New roads will usually require street lighting, in which case electricity will be required and if the new road leads to an estate, the utilities companies will normally lay their cables, ducts and supplies beneath the footway.

The provision of new supplies needs to be considered during the early stages of the development so that the groundworks such as pipe crossings, ducts and drawpits can be co-ordinated and programmed for construction alongside the main works.

3.8 Police and emergency services

When major works are to be carried out in a highway, for example the laying of new mains, services and structural carriageway repairs, the works often have significant effects on the local traffic conditions and in busy cities the effects of these works can be widespread.

It will often be found that the emergency services have preferred routes for access and consultations with these services will establish if the proposals are likely to affect those routes.

Minor works such as a single connection to a public sewer under the highway are not generally considered to cause significant disruption and although traffic management will be necessary, it is unlikely to be on the scale of that required for a junction improvement scheme or for the construction of an interchange.

Access for the emergency services – police, fire and ambulance – should not be impeded for obvious reasons and it is therefore advisable to undertake consultations with these bodies, especially when considering traffic management proposals, so that their local knowledge especially with regard to local traffic conditions and patterns can be considered and alternative routes for their vehicles planned if necessary.

3.9 Others

Other consultees may include providers of public transport such as bus and tram operators. As with the utilities and stats companies, the same two reasons for consulting with these organisations apply, although the reasons are more likely to be as a result of a major highways scheme or the implementation of traffic calming measures. Either of these types of project may necessitate road closures or the diversion/re-routing of such services, and in some instances for public safety it may be necessary to relocate bus stops for the duration of the works.

4

Environmental issues

4.1 Introduction

As long ago as 1972, the first international conference concerning the environment was held in Stockholm, Sweden, but it was to take another fifteen years before world leaders were told in no uncertain terms that there would be severe consequences if the pollution and depletion of natural resources was to go unchecked. It was suggested that for development to be sustainable it should meet the needs the present generation without compromising the needs of the future.

Since 1988 the campaigns for environmental awareness have gained impetus and influence and the possible effects of construction projects on the environment are finally starting to be considered in the early stages of their development.

To chart the progress, then, of environmental strategies in the UK since the Stockholm conference in 1972, the first major publication came about in 1987 with a report published by the United Nations entitled *'Our Common Future'*.

In 1988, the then prime minister, Margaret Thatcher gave a speech to the Royal Society which gave a well needed boost to the environmental movement in the UK and shortly after, the Green Party gained approximately 15% of the vote in the 1989 European elections.

In 1990, the government published a white paper entitled *'This Common Inheritance'* and this was the first comprehensive environmental strategy. Also in 1990, the IPCC (the Intergovernmental Panel on Climate Change) published its first report which predicted, as a result of global warming, an average rise in sea levels of 60mm every ten years and an average global temperature rise of 1°C by the year 2025.

Two years later in 1992 the *'Earth Summit'* was held in Rio de Janeiro and it was there that over 120 heads of state formulated plans to reduce the harm being caused to the environment and adopted an action plan to achieve worldwide sustainable development.

This action plan, known as Agenda 21, required full national participation by each of the countries involved in the formation of effective strategies for sustainable development.

As a follow up to the agreements at the earth summit, in 1993 the Department of the Environment issued a consultative paper entitled '*UK Strategy for Sustainable Development*' and this was followed up in 1994 by papers on climate change, sustainable development, biodiversity and forestry.

In April 1996, under the Environment Act 1995, the Environment Agency was formed with its principal aims being:

- continuing improvement in the quality of land, air and water
- the active encouragement of the conservation of natural resources
- to protect and warn against flooding
- to achieve reductions in waste through reuse, recycling, improved disposal standards and minimisation
- to conserve and enhance inland and coastal waters and their use for recreation
- to improve and develop salmon and freshwater fisheries
- to secure the remediation of contaminated land
- to manage water resources
- to balance water resources between environmental needs and abstraction

It can be seen then, that over recent years, changes in law and general opinion have brought about fundamental changes in the approach to design.

Environmental protection is not a new concept, but the development of scientific knowledge and the awareness of environmental issues combined with pressure and policy from the European Union have led to the implementation of new concepts in controlling the way pollution affects the environment.

Statute and common law provide the roots for most of the current UK environmental legislation, whereas in the European Union the legal system is based upon written codes, regulations and directives.

In the UK, an integrated pollution prevention and control (IPPC) policy has been in use for a number of years, but in other EU states, the general practice is to consider separately the emissions into air, water and land. IPPC is currently being absorbed into the EU and is likely to be adopted by the European Parliament.

For certain types of development, assessments of the likely environmental impact can be required and depending upon the category of the development as defined by the Town and Country Planning (Assessment of Environmental Effects) Regulations 1988 (as amended) such an assessment is either compulsory or at the discretion of the local planning authority. The Environmental Impact Assessment, as this requirement is known, has a key role in all types of development.

The development of contaminated sites and concerns over the risks associated with development of contaminated land have increased steadily over recent years, but it is not only contaminated sites which present significant threats to the environment. All developments should undergo some form of assessment of their impact on the environment and as a minimum when an Environmental Statement is required, must or should consider the following points:

- a description of the proposed development (mandatory)
- the data necessary to identify and assess the principal effects of that development on the environment (mandatory)
- a description of the likely significant effects on the existing environment (mandatory)
- a description of any measures proposed to counteract any adverse effects (mandatory)
- a non-technical summary of the first four points above (mandatory)
- a description of the processes involved which will arise from or during the life of the development, from the construction phase including any temporary works through to final usage and likely operations within the developed site. In the first part this could include subjects such as the excavation of borrow pits or quarries and the control of groundwater/dewatering, and should consider

emissions, trade effluents and waste products. All stages should consider the likely direct and indirect effects on the environment
- information relating to the materials proposed for use in both temporary and permanent works and details of any possible adverse effects arising from the production processes involved in obtaining those materials
- principal alternatives that have been considered for the development
- substantiation as to why the proposed development, materials, and intended nature of the development are considered to be the most environmentally preferable
- characteristics of the development

The Environmental Protection Act 1990 introduced the system of integrated pollution control (IPC) and this was aimed at improving the regulation of industrial processes. It is regulated by the Environment Agency. The Environmental Protection (Prescribed Processes and Substances) Regulations 1991 identify six categories of processes as follows:

- fuel production processes
- metal production and processing
- mineral industries
- chemical processes
- waste disposal and recycling
- other industries

Part I of the Environmental Protection Act 1990 makes it an offence for any person to carry out a prescribed process other than in a manner set out in an authorisation granted by an enforcing authority (i.e. an IPC authorisation).

Normal construction activities as such are not designated as prescribed processes and will therefore generally fall outside the system of integrated pollution control. It should be noted however, that with the development of Design, Build, Finance and Operate (DBFO) and Private Finance Initiative

(PFI) schemes this may mean that construction companies, as part of their involvement in operating schemes which carry out prescribed processes, become involved in the integrated pollution control system.

4.2 Planning stage

Where a proposed development has the potential to have a detrimental effect on the environment either during construction or post-construction, certain pollution control regimes may apply.

These regimes are separate from planning control but are intended to complement the planning process and planning controls are not intended as a means of regulating polluting activities. The Department of the Environment has published PPG23 Planning and Pollution Control to define the boundaries between the regimes and to remove the potential for duplication of statutory controls.

PPG14 is probably one of the most relevant sources of guidance for developments on contaminated land and has as its main concern, the development of land which has the potential to cause pollution or contamination or is otherwise unstable or potentially so.

Planning authorities can exercise control over potentially polluting activities when they are applied to the storage of hazardous substances, in which case the Planning (Hazardous Substances) Act 1990 and the Planning (Hazardous Substances) Regulations 1992 come into effect; but it should be noted that the relevance of these pieces of legislation is probably limited to the storage of large quantities of hazardous substances.

Where separate applications for planning permission and authorisation under integrated pollution control are required, early consultation and liaison between the planning authority and the pollution control authority will enable each body to assess the impact of the scheme.

Further, it will speed up the process if each has access to the same information at the same time. It should be noted, however, that integrated pollution control authorisation will not be granted until planning permission has been obtained.

The appropriate information may not always be available at the time the applications are made and in such cases, staged applications may be permitted for integrated pollution control authorisation. This allows discussion between the parties involved which may smooth the progress of such applications.

As we have already seen, certain types of development require an Environmental Impact Assessment (EIA) and this will be considered by the local planning authority in determining an application.

There are two categories of development: Schedule 1 – for which an EIA is compulsory and Schedule 2 – for which an EIA is requested at the discretion of the local planning authority, and these are brought about by the Town and Country Planning (Assessment of Environmental Effects) Regulations 1988. The regulations stipulate for which types of development an Environmental Impact Assessment is required.

Schedule 1 projects include schemes which can have a significant environmental impact such as infrastructure developments, major highways schemes, power stations, chemical process plants and waste disposal installations.

Schedule 2 projects include schemes on a lesser scale such as mineral extraction, holiday villages, water and waste water treatment installations and food manufacturing plants.

Guidance on what projects constitute a significant effect on the environment is set down in Department of Environment Circular 15/88: Assessment of Environmental Effects and suggests that an EIA is required:

- where the size and scale of the project is of more than local significance
- where the location of the development is potentially sensitive, such as a site of special scientific interest (SSSI) or in a national park
- where complex or adverse polluting effects may occur as a result of the development

The circular also gives guidance on the categories of Schedule 2 projects and although each Schedule 2 project is to be assessed by the local planning authority on the project's individual circumstances, the guidance relates to the size and scale of the project and the area of land where the project is located.

In addition to the circular, recent amendments in European legislation have brought about a widening in the scope of UK legislation which covers Environmental Impact Assessments and which came into force before March 1999. Of particular relevance to the construction industry are:

- Schedule 1 projects now include developments associated with the nuclear industry and landfill sites
- Schedule 2 categories have been clarified
- local planning authorities can now be required to give an opinion as to the information which should be included in an Environmental Statement
- the Environmental Impact Assessment should include an outline of the alternatives which have been considered for the project and should give an indication of the reasoning behind the choices made
- Schedule 2 projects are now subject to a more formal process to determine whether or not assessment is required
- public consultations are now considered by the local planning authority in its determination of an application and there is now a requirement that any reasons for its decisions should be made public

When considering an application for a project where an Environmental Impact Assessment is required, the local planning authority must consider at least the following two items:

- the contents of any Environmental Statement
- any representations made by:
  - the statutory consultees and/or
  - any other person concerning the likely environmental effects of the proposed development

Further guidance on Environmental Assessment can be found in Department of Environment Circular 3/95: Permitted Development & Environmental Assessment. This document sets out procedures which should be followed in the preparation of an Environmental Impact Assessment.

4.3 Construction and waste materials

The very nature of construction means that large quantities of waste can be generated, whether this is in the form of wasted materials or brought about by excavations which give rise to the requirement for the disposal of large quantities of materials.

The subsequent transportation and disposal of excavated materials off site will incur financial penalties in the form of landfill tax which is levied on the disposal of most waste.

Waste can be defined as any substance or object which is intended to be discarded and is legally defined in the Waste Management Licensing Regulations 1994. In the UK, such waste is referred to as 'Directive Waste'.

For the construction industry there are probably two criteria which determine what constitutes 'Directive Waste', but if in doubt, a contractor should assume that all waste will be classified as 'Directive Waste' until advised otherwise by a regulator. The definition of 'Directive Waste' is:

- material which is intended to be discarded as defined by the person in possession of the material. This includes material which is so classified and will be treated as waste by the regulating bodies even if it is to be used as a raw material by a third party

There are also two instances where waste will not be classified as 'Directive Waste' and these are:

- where the waste is to be transported from site to be processed off-site and then returned to site in its processed condition
- where, under the terms of a contract, the waste is to be re-used in the site from which it originated

In addition to generating waste, the construction industry is also recognised for its innovation in re-using waste and recycling materials, particularly in the construction of roads.

When dealing with waste, the Environmental Protection Act 1990 is particularly relevant to contractors on site as this legislation imposes a duty of care on all those who deal with waste.

This duty is to ensure that all those involved with the production, handling, transport and ultimate disposal of waste deal with the waste in a responsible manner. Section 34 of the Act applies to anyone who 'imports, produces, carries, keeps or disposes of controlled waste...' and imposes a duty on those people to ensure that:

- the waste is contained (i.e. it does not escape)
- the waste is transferred only to an authorised person
- a written description of the waste accompanies its transfer to another person
- no other person disposes of the waste in a manner harmful to the environment or injurious to human health; or in an unlawful manner

A further publication, the Department of Environment document 'Waste Management: The Duty of Care: A Code of Practice' was first published in 1996 and provides guidance on the handling and disposal of waste. It is also 'admissible in evidence', which means that any court dealing with a prosecution for failing to comply with the duty of care must take into account the relevant provisions of the code of practice. For contractors, this means that:

- they must store, contain and handle all waste properly
- they must only deal with licensed operators and carriers and provide transfer notes for the waste
- they must take steps to ensure correct disposal
- they must provide a description of the waste

Any contractor transporting material off-site for disposal to a landfill site should be aware that the materials being transported will probably be subject to the landfill tax. This tax was introduced in the 1994 November budget as an incentive to deter the landfilling of waste and to encourage recycling and is applied per tonne of waste, with the rate of tax per tonne depending upon the classification of the waste (i.e. inert or non-inert)

Even where materials are not classed as 'Directive Waste' or are exempted from being classed as such, the materials should be handled with care especially when there is the potential for contamination of the soil, groundwater pollution or other emissions which may give rise to action under any of the previously mentioned legislation.

4.4 Site hazards

It has been said that a contaminated site does not represent a threat to the environment until it is disturbed. As long as the contamination is contained wholly within the site e.g. the site is not leaking toxic waste in its undisturbed condition, then this can be true.

Once construction works start, however, there is considerable scope for the activities to bring about environmental issues and liabilities. Any construction works should be undertaken at minimal risk to the environment.

The risks are not restricted to works in the ground which may affect, for instance, habitat, groundwater and watercourses, but may extend to include noise and air quality (dust and smell for example) and it is these last two which can be often overlooked.

Although it is unlikely that any claim in common law brought about by an aggrieved third party for the noise, vibration and dust caused, for example, by a morning spent by a contractor in breaking out a concrete slab, would be successful (such works are transient and the nuisance is temporary), if it could be proved that the works gave rise to a statutory nuisance as defined by the Environmental Protection Act 1990, then action could be brought against the perpetrator. In relation to dust, noise and smell, statutory nuisance can be defined as:

- noise, smoke, fumes, gas, dust, steam, smell or other effluvia arising... and being prejudicial to health or a nuisance
- any accumulation or deposit arising... so as to be prejudicial to health or a nuisance

There are several common sources of pollution and potential threats to the environment which arise from construction activities and to avoid breaches of legislation contractors should consider at least the following:

- contractor to satisfy himself as to the neutrality of the land (i.e. contamination not present)
- the potential to pollute soil, groundwater and watercourses either directly or indirectly, e.g. by spillage (directly) or by damage to a foul sewer which leads to pollution of a river (indirectly)
- the potential for a lack of evidence of contaminants at the surface to be misleading. Sub-surface conditions may be completely different
- the potential to pollute potable water supplies by excavation of or damage to unidentified water mains either in contaminated or non-contaminated land
- the possible existence of isolated pockets of contaminated land
- the pollution potential of materials removed from or brought to site including their methods of removal/transport
- the storage of hazardous substances

In addition to any hazardous substances which may be present on site, contractors can introduce their own contaminants during the construction process and these can include paints, solvents, cement, plaster and fuels for machinery.

Hazardous substances are subject to their own legislation, the Control of Substances Hazardous to Health Regulations 1994 (more generally referred to as the COSHH regulations), which impose duties upon contractors and employers in general to minimise the risk imposed by such substances and which include a requirement for assessment of the operations which use such substances.

Contractors and site operatives should be aware that there is a need to control hazardous substances and to carry out any operations which have the potential to cause pollution within a clearly defined procedure.

4.5 Water and land drainage

In terms of the frequency of prosecutions under legislation surrounding water pollution, the construction industry is the most persistent offender. The probable causes for this are that risks arise wherever construction activities are carried out adjacent to watercourses, rivers, streams and estuaries, or where underground aquifers are close to the works.

Nationwide, there are adequate water resources, although locally this may not always be the case, especially during long periods of dry weather where population densities are high.

Options have been considered to overcome these problems, such as the transfer of water from areas with a surplus to areas where water levels have fallen to dramatically low levels.

For water supply to be sustainable, however, the long term strategies need to consider effective management of water demand, leakage control and regional/local pipelines for the transfer of water.

The Environment Agency is responsible for the management of water resources and the protection of the aquatic environment under the Water Resources Act 1991 which covers inland, coastal, estuarial and ground waters.

4.6 Contaminated land

In 1988, a study by the Department of Environment suggested that up to 27,000 hectares of land in the UK could be contaminated and traditionally, the remediation of such sites has involved either off-site disposal of the contaminated materials or the encapsulation of the materials on site.

Current trends are towards redevelopment of existing or 'brownfield' sites rather than usage of new or 'greenfield' sites and the costs of clean-up can be significant. There are several factors which have led to this move towards utilisation of existing sites and clean-up operations:

- public pressure. Greater public awareness has led to demands by the public for a cleaner environment
- regulatory pressure. Better enforcement of existing legislation and the

creation of new legislation have brought about more stringent control
- large multi-national companies' adoption of worldwide environmental standards. This course of actions has also given those companies an environmentally responsible profile
- economics. Clean-up costs can be significant and organisations cannot afford to cause pollution
- environmental liabilities associated with land transactions and the acquisitions of other companies have led to those companies being more responsible for their actions

Contaminated sites require careful consideration if pollution is to be avoided, and planning authorities seek to encourage developers to utilise brownfield sites. In developing brownfield and contaminated sites, there are inherent risks in unforeseen ground conditions and clean-up operations and there may be liabilities arising from ownership or occupation of such land.

Notwithstanding these risks, however, planning authorities want to ensure that no development gives rise to risks to health or to the environment and, as we have already seen, can and will impose conditions as part of granting planning permission that acceptable and appropriate measures are taken to ensure that such risks or pollution do not occur.

The nature and extents of clean-up operations are chiefly outside the scope of this book but can be summarised as follows:

- capping of contaminated soil. Capping reduces the amount of precipitation that infiltrates into the contaminated soil and prevents direct contact with the contaminated soil by humans and wildlife
- excavation and removal of contaminated soil. Excavated contaminated soil is usually taken off-site to a landfill site or incinerated
- soil stabilisation/solidification. This involves the mixing of cementitious materials or pozzolanic reagents with the contaminated soil to create a chemically inert and stable matrix. It can be carried out by ploughing in the additives to the contaminated soil or by mixing above ground on site.

- low temperature thermal stripping. In this application, the soil is disturbed, often by ploughing or excavation, and heat is applied sufficient to drive off volatile contaminants.
- biological treatment. This method of treatment utilises bacteria to degrade the organic pollutants present in the soil. The bacteria may already be present in the soil, in which case enhanced bacterial growth may be encouraged by the addition of nutrients, or the bacteria may be introduced into the soil.
- vacuum extraction. Unsaturated soils contaminated with volatile and semi-volatile organic compounds can be cleaned up using this technique which uses a system of vacuum wells installed throughout the contaminated area which withdraw air from the soil and pass it through a filter.
- in-situ vitrification. This technique requires a series of electrodes to be placed in the ground to melt inorganic contaminants. A current is applied which raises the temperature in the area enclosed by the electrodes to about 1600°C and the molten contaminants can be left in place.
- pump and treat: This is a treatment for contaminated groundwater and involves the installation of groundwater recovery wells in the locality of the contaminants to extract the groundwater to an above ground treatment plant.
- point of use treatment. This is a type of groundwater clean-up operation designed for use adjacent to abstraction wells which abstract water for drinking purposes or agricultural use and remove the contaminants prior to the distribution of the abstracted water.

5

Drainage design

5.1 Introduction

Many forms of drainage systems have existed since ancient times. Most of the early drainage systems were for disposal of surface water only, with foul waste generally being disposed of in the most convenient manner, usually via a hole in the ground, ditch or midden.

More recently, the physical design of sewers has progressed and from an early approach which was based almost entirely on experience, developments in surface water sewerage evolved and incorporated a period when almost all surface water drainage design was based upon theory. This in turn led to research being undertaken, which indicated that the rules of thumb employed by early drainage engineers were surprisingly accurate.

Current practice is generally to design separate systems to carry foul and surface water run-off, but many of the sewerage systems in existence today have developed from systems which were originally designed to take surface water only.

In the middle of the 19th century, the invention and subsequent widespread use of the water flushed toilet led to the construction of cesspits which were connected to the surface water sewers and many of the receiving sewers found themselves conveying additional waste. The industrial revolution also contributed substantially to the additional flows the surface water sewers were beginning to carry and led to an increase in the amount of industrial flows entering the systems.

As local populations grew and the requirement for paving of roads and paths increased, this naturally led to further flows entering the sewers and to problems such as flooding and surcharging, with most of the sewer systems becoming greatly overloaded.

In the infancy of drainage design, there had been no strategy to allow for future growth or development and with the additional flows generated by industry and an expanding population, drainage engineers were faced with solving immediate problems rather than providing long term solutions.

The sewer systems thus developed in a haphazard fashion, with overflows and complex cross-connections being provided to relieve the overloaded sections of sewer by diverting some of the excess flows into sewers which at the time had spare capacity. Currently, many of these early combined systems are still in use, although most of the combined flows have been separated to foul and surface water sewers.

Currently, surface water sewers are provided to convey the run-off from roofs and paved areas such as car parks and roads to a suitable point of discharge but their design is not necessarily an accurate science. Many of the parameters and factors used are best estimates, for example, predicted population growth, predicted water consumption or predicted rainfall. The design storm is a convenient way of representing, in terms of a rainfall event, what can be considered in equivalent structural engineering terms as a design load. The rainfall event will impose design loads on the drainage system in much the same way as a design load will impose a load on a structure.

The design storm will have a return period and duration, but it should be remembered that although a storm may be classed as having a five year return period, this only means that statistically such a storm is likely to occur x times over a much longer period, for example twenty times every hundred years. It could be, then, that a five year storm could occur twice in one year.

5.2 Terminology

The following definitions apply to the terms commonly encountered in drainage practice:

Benching: The sloping raised base of a manhole to the sides of and above the channel. Normally topped with a smooth granolithic concrete

Combined sewage: The discharge which includes surface water run-off and foul water in the same conduit

Combined sewer: Sewer constructed to convey foul water and surface water in the same conduit.

Conduit: A duct used for the conveyance of sewage e.g. pipes, channels, box culverts, etc.

Crown: The topmost point of the external surface of a circular or oval section conduit

Dry weather flow (DWF): The flows encountered in a sewerage system under normal dry conditions. May also be the rate of flow averaged over a twenty-four hour period

Dry well: The underground chamber used to house pumping equipment

Foul sewer: A sewer constructed to convey foul water

Foul sewage: The discharge from appliances such as toilets, sinks, baths, etc.

Gravity sewer: A sewer in which the sewage flows under a natural hydraulic gradient and is not pumped

Hydrograph: A representation of a rainfall event, where time and rainfall are plotted graphically against each other

Invert: The lowest point on the inside surface of a conduit

Inverted syphon: A section of pipe in which the sewage flows under pressure due to the pipe being below the hydraulic gradient

Lamphole: A small shaft on a sewer used for the purpose of lowering lamps into the sewer to enable inspections to be carried out

Manhole: A chamber constructed around a conduit for the purposes of providing access into the conduit from ground level and where there is a change of conduit size or shape, direction or gradient

Pumping main: A sewer in which the sewage is pumped under pressure to a convenient point of discharge

Pumping station: Point on a sewerage system where the sewage is collected and from where it is pumped through a rising main to a higher level

Sewage: The liquid conveyed in a conduit. May be surface water run-off, foul water or trade effluent

Sewerage: A system of pipes and conduits laid for the purposes of conveying sewage

Soakaway: Conventional soakaways may be a pit, trench or chamber constructed below ground at the end of a surface water drain to allow dispersal of surface water directly into the ground. Pits and trenches may be filled with granular free draining material. Chambers are perforated. Borehole (or deep bore) soakaways are constructed with a perforated liner which is sunk to a predetermined depth (e.g. to reach fissured rock or permeable strata)

Soffit: The topmost point of the inside face of a conduit

Storm water: Foul sewage diluted by surface water run-off

Storm water overflow: A device such as a weir or orifice situated within a combined sewer to allow diversion for discharge elsewhere of storm water which is above a pass forward rate agreed by the Environment Agency

Surface water: The run-off from roofed and paved areas arising from a storm event

Surface water sewer: A conduit constructed for the purposes of conveying surface water

Trade effluent: A discharge produced in the course of any trade or industry which excludes domestic sewage

Wet well: The underground chamber of a sewage pumping station which receives the sewage.

5.3 Layout of sewers

There are many factors which must be considered when determining the route or position of a sewer.

In general it is preferable to construct sewers which are going to remain in private ownership within the private areas of a new development.

Sewers which are going to be offered for adoption, however, i.e. those which will be maintained by the drainage authority, should be located within public areas e.g. roads, wherever practicable, although easements may be granted and will be required where these sewers pass through land in private ownership.

It will always be preferable to lay public sewers in roads rather than footways, although sewers laid under verges are more acceptable than those in footway locations.

Individual drainage authorities are likely to have their own specifications for particular elements of drainage works and these may stipulate items such as minimum clear openings for manholes or minimum distances from kerblines to pipes and manholes.

When routing pipes, consideration should also be given to the proximity of existing services such as gas mains, other service pipes and sewers, ditches etc. and also other buildings, whether existing or proposed.

Investigations should be carried out into the proximity of other services by liaison with other statutory bodies such as cable TV, telecommunications, electricity, gas and water.

Maps showing the locations of the underground apparatus are generally available for most areas from most of the utility companies, although the accuracy of these maps can never be guaranteed and it is essential that prior to any excavations being carried out methods such as the use of cable avoidance tools and signal generators (CAT and Genny) or the careful excavation of trial pits are employed to mark out on the ground the exact positions of any underground services.

For economic designs, consideration should be given to the laying of foul and surface water sewers in the same trench, although where the difference in level of each sewer is significant, there may not be a significant cost saving and the long term integrity of the pipe at the higher level may be prejudiced if inadequate support is provided during and after construction. Where wide trenches are required due to structural design of the pipelines, however, combined trenches may not be particularly economical in construction.

Town planning schemes or development plans will provide valuable information on the areas designated for development and would be incomplete without their having given consideration to the sewerage of any new development.

Materials and methods of construction should take into account factors such as ground conditions, intended methods of pipelaying, practicability of construction and the nature of the sewage.

The safety of the access should also be considered when siting manholes, particularly if the manhole will be located in a busy junction or similar. Safety in construction as well as in use should be considered.

Depths and gradients should be such that the design represents an economic and practical solution to the problem of drainage. Generally a gravity system will fulfil these criteria, but certain factors may determine that a sewer needs to be pumped and these may include the need to avoid excessive depths of sewer or to drain low areas of a site, the need to overcome an obstacle or to centralise sewage disposal or the need to provide an outfall when a gravity connection is not available.

Sewers should be laid at such depths that there is adequate cover to the crown of the pipe or where this is not possible, the pipe or conduit should be surrounded in a concrete bed of sufficient strength to carry the loads likely to be imposed upon the pipe. The depth of sewer should also take into consideration the likelihood of future development so that any new connections can also be made economically.

Backdrop manholes are frequently used to economise the construction of the drainage system especially where the site topography varies frequently in level. Other locations where backdrop manholes may be appropriate include where a minor sewer connects to a major sewer and the major sewer is at a significantly lower level, where abrupt deepening of the sewer is required to overcome an obstacle such as under a railway or a river. In the past backdrop manholes were constructed to dissipate energy and to reduce flow rates to prevent scour.

Ventilation is not usually required in gravity sewers except at the head of the system and where access is required such as in wet wells, in which case ducts are usually laid close to ground level from the wet well to a convenient point. The ducts normally terminate 1.8m to 2.4m above ground level. Ventilation at the head of the system is normally provided by the use of vent pipes and air admittance valves in the private section of the sewer as required by current building regulations.

Ventilation should, however, be provided in locations where the sewer is likely to become tidelocked or where the sewer has a very flat gradient as this measure will allow air to escape as the sewer becomes charged particularly during high tides and intense storms and will prevent the pipes from becoming airlocked. This situation can also occur where the main trunk sewer is surcharged and flow backs up or cannot discharge from the tributary sewer.

On pumped systems, air release valves must be provided at any deviation between the the physical gradient and the hydraulic gradient and at distances not greater than specified in the relevant British Standard. At low points, washouts or flushing chambers are normally provided.

5.4 Hydraulic design of sewers

Once the predicted flows have been calculated for a particular catchment, the next step is to ensure that the proposed network of pipes will have adequate capacity to carry the flows.

Surface water sewers generally convey only a small percentage of suspended materials, but the grit loadings can sometimes be very high, especially during 'first flush' which occurs after a dry period when all the silt, debris and grit that has accumulated on the surface is carried into the sewers with the run-off.

Initially, the layout for the proposed sewer network should be drawn at a scale sufficient to cover the whole catchment on one drawing. Points of discharge should be identified (e.g. existing watercourses or public sewers) and the proposed sewers should generally follow the ground contours to reach these points.

From this layout, longitudinal sections along the route of the proposed sewers can be drawn and this will assist the designer in determining gradients and diameters as well as ensuring that excessive lengths of deep excavation and pipes with inadequate cover can be avoided.

5.4.1 Flow and energy losses

Much research has been undertaken into the capacities of different types of conduit, from small diameter pipes through to large open channels. The volume of run-off which can be conveyed by a conduit depends upon three factors: gradient, energy losses and conduit size.

Gradients should be selected to provide at minimum, self-cleansing velocity, although in practice, existing ground levels and contours will often dictate what gradients can be achieved. When considering pipe gradients, it is general practice to assume normal flow conditions (i.e. no surcharging) to determine the hydraulic gradient.

Under these flow conditions, the hydraulic gradient will generally be parallel to the line of the pipes. When the pipe is empty, its invert is usually accepted as representing the hydraulic gradient of the sewer. When the pipe is full, however, the soffit of the pipeline represents hydraulic gradient. Pipes should therefore be laid with soffits level rather than inverts level.

For example, if a manhole has an incoming pipe with a diameter of 375mm and an outgoing pipe with a diameter of 450mm, the invert of the outgoing pipe should be 75mm lower than that of the incoming pipe.

The designer should remember that flow velocity is dependent upon the rate of flow and not the pipe diameter. This is often overlooked and is a problem often compounded by a proposed pipe being close to its maximum capacity when the hydraulic gradient between the upstream and downstream ends is relatively flat under normal conditions.

In this instance, there will usually be a difficulty in achieving self cleansing velocity in the specified pipe, but the specification of a larger pipe at a flatter gradient will not overcome this difficulty when the larger pipe is running in the part full condition.

The smaller pipe laid at the same gradient will achieve better self cleansing velocity but will be overloaded, resulting in surcharge. This surcharge will ultimately increase the hydraulic gradient which will in turn increase the velocity in the pipe and thus mean that the velocity required for self cleansing is approached.

For determining gradients, significant use was made in the past of McGuire's Rule which suggested that 150mm (6 inch) pipes should be laid no flatter than 1 in 60 and that 225mm (9 inch) pipes should be laid no flatter than 1 in 90.

As a general rule of thumb, McGuire's Rule can still be used, but for final design it is not recommended as the outcome will produce an uneconomical scheme, with very shallow sections of sewer at the head of the network, excessively steep gradients and over-sized pipes. Further, upsizing the pipe to slacken the required gradient is no good if there is insufficient flow in the pipe.

To determine accurately the capacity of a conduit at a given gradient, there are several formulae in regular use, and some of these are listed in the later section in this chapter. Alternatively, the publication *'Tables for the hydraulic design of pipes, sewers and channels'* tabulates capacity and discharge for full and part full conduits at given gradients and roughness factors and from which values can be read directly and accurately.

It is usual to design surface water sewers so that self cleansing velocity is achieved when the pipe is running full. This is rarely the case with surface water sewers, so self cleansing velocity is normally taken as 1 metre/second for pipe full conditions for pipes up to 900mm diameter. For pipes larger than this, the self cleansing velocity should be a *minimum* of 1 metre/second.

Several formulae have also been developed to calculate energy losses (also known as head or pressure losses) in pipes. The Darcy-Weisbach equation (developed around 1850) is one of the most commonly used formulae and demonstrates that energy losses are dependent upon four factors as follows:

- Velocity. When the velocity is doubled, the energy losses increase by four times that amount. Headloss is proportional to the square of the mean velocity
- Length. Length has a direct effect on energy losses as the longer the conduit, the higher the losses

- Friction
- Size. The diameter of a pipe has the most significant effect on energy losses. The smaller the pipe, the higher the energy losses. If a pipe diameter is reduced from 300mm to 150mm (i.e. by half), then the energy losses are increased by 32 times

The Darcy-Weisbach equation takes the form:

$$h_f = \frac{\lambda l v^2}{2gd}$$

where

h_f = headloss due to friction
λ = friction factor (referred to as f in American practice or $4f$ in early UK practice)
l = pipe length in metres
v = velocity (metres per second)
g = gravitational acceleration
d = pipe diameter (metres)

λ can be calculated from the following formula known as the Blasius equation

$$\lambda = \frac{0.316}{R^{0.25}}$$

where

R = Reynolds number up to 10^5

λ was investigated by Blasius in 1913 and later by Nikuradse in the early 1930s. Blasius developed the formula for λ shown above and discovered that the roughness of a pipe has no effect on friction and that λ is dependent upon the Reynolds number. The later experiments by Nikuradse, however, drew the opposite conclusion and found that λ depended only on the roughness of the pipe and was independent of the

Reynolds number.

It can be shown graphically that both were in fact correct, and that λ varies with Reynolds number and pipe roughness. The Blasius experiments were carried out where the flows had relatively low Reynolds numbers (up to 100000). Nikuradse's experiments however, were carried out where the flows had considerably higher Reynolds numbers (100000+).

For Reynolds numbers greater than 10^5, λ can be calculated from the formula:

$$\frac{1}{\lambda^{0.5}} = -2\log\left[\frac{k_s}{3.7d} + \frac{5.1286}{R^{0.89}}\right]$$

where

k_s = roughness value (mm)
d = pipe diameter (metres)
R = Reynolds number greater than 10^5

This formula was developed by Barr in 1975 and gives a solution for S_f to an accuracy of $< \pm 1\%$. (S_f = hydraulic gradient h_f/L).

Since there were two correct sets of results for the Blasius and Nikuradse experiments, it followed that there were two types of flow being investigated and that at some point λ would depend upon both the Reynolds number *and* the pipe roughness. This transition zone was investigated in the 1930s by Colebrook and White and they developed a formula to cover flows within that zone.

The Colebrook-White transition formula combines the von Karman-Prandtl rough and smooth laws and is written as follows:

$$v = -2(2gds)^{0.5}\log\left[\frac{k_s}{3.7d} + \frac{2.51v}{d(2gds)^{0.5}}\right]$$

where

v = velocity (metres/second)
g = gravitational acceleration
s = hydraulic gradient
k_s = effective roughness (mm)
d = pipe diameter (metres)

Typical pipe roughness (k_s) values can be taken from ***Table 1*** or from manufacturers' product literature, but the designer should be aware that any values for k_s put forward by manufacturers are for new pipes and the figures do not necessarily take into account slime, grit or other deposits which accumulate in the conduit. When selecting k_s values, the designer should also consider quality of workmanship such as pipe jointing.

In 1963, Hydraulics Research Papers suggested k_s values based upon the condition of the sewer of 0.6mm (good), 1.5mm (normal) or 3.0mm (poor). In 1964, Ackers suggested that the values of k_s in the Colebrook-White formula took into account the build up of slime in pipes constructed of any material as the k_s value of 1.5mm over the bottom 25% of the pipe surface is equivalent to a Manning's n-value of 0.012.

Current practice is to design surface water sewers with a k_s value of 0.6mm. This is specified in the current edition of *'Sewers for Adoption'* and is a requirement should the sewers be offered for adoption.

Manning's formula for open channel flow can be used in the following form for pipes running either full or half full, when the equation becomes:

$$v = \frac{0.004\, d^{2/3}\, I^{1/2}}{n}$$

where

v = velocity in metres per second
n = friction coefficient
d = diameter of pipe in millimetres
I = pipe slope

Pipe material	Typical k_s values (mm)		
	Good	Normal	Poor
UPVC	0.06	0.06	0.06
Precast concrete	0.06	0.15	0.6
Clayware, sleeved joints	0.03	0.06	0.15
Clayware, spigot & socket joints (*)	0.03-0.06		
Brickwork (**)	1.5	3.0	6.0
Rising main – 1m/s mean velocity	0.15	0.3	0.6
Rising main – 1.5m/s mean velocity	0.06	0.15	0.3
Rising main – 2m/s mean velocity	0.03	0.06	0.15
Earth channel – straight, artificial	15	60	150
Earth channel – straight, natural	150	300	600

(the higher values of k_s should be used for clayware pipes greater than 150mm diameter; ** for brickwork in well pointed condition)*

Table 1 *Typical pipe roughness (k_s) values*

The original Manning formula, presented in 1889 to the Institution of Civil Engineers of Ireland, was developed from about seven other formulae verified by observations. It has become simplified to:

$$C = \frac{R^{1/6}}{n}$$

where

C = Chezy coefficient
R = hydraulic radius

For pipe flows, the Crimp and Bruges formulae are

$$v = 0.33\, d^{2/3} i^{1/2}$$

$$Q = 26 \times 10^{-8}\, d^{8/3} i^{1/2}$$

when n has a value of 0.012 and where

v = velocity (metres/second)
Q = discharge (cubic metres/second)
d = pipe diameter (metres)
I = pipe gradient

The Crimp and Bruges formulae are based upon a Manning's *n* value of 0.012, a value which has been found to be acceptable for sewers in current use, and the formula can be used satisfactorily for existing sewers constructed from most pipe materials. Further *n* values are set out in ***Table 2*** and should be used in for conduits over 900mm diameter.

Type of conduit	*Manning's n*
Concrete lined canals	0.012–0.017
Rough masonry	0.017–0.030
Rough earth canals	0.025–0.033
Smooth earth canals	0.017–0.025
Cast iron pipes	0.013–0.017
Timber (planed)	0.008
Brickwork and stonework – old	0.020
Brickwork and stonework – fair	0.017
Brickwork and stonework – good	0.015
Concrete	0.015
Corrugated metal culverts	0.021

Table 2 *n values for use with Manning's formula*

5.4.2 Part full pipes

When a sewer is designed to run at less than pipe full, difficulties can arise in achieving self cleansing velocity. ***Table 3*** gives the relationship between proportional depths, proportional discharges and proportional velocities. The proportional depth is given by:

$$d_p = \frac{\text{depth of flow}}{\text{pipe diameter}}$$

where the depth of flow and pipe diameter are measured in the same units.

To obtain the proportional velocity from Table 3, read across from the appropriate proportional depth and multiply the full bore velocity by the appropriate factor in the table to calculate the proportional velocity. The proportional discharge can be calculated in much the same way. The table can be read in any direction, starting with a known proportional discharge, for example to obtain proportional velocity.

As a worked example, let's take a flow of 30 litres/second flowing down a 300mm diameter pipe laid at 1 in 250 gradient. The full bore capacity and gradient of that pipe are 69.856 litres/second and 0.988 metres/second respectively. The proportional discharge is therefore:

$$Q_{prop} = \frac{30}{69.856} = 0.429$$

Interpolating from the table gives a factor for proportional depth of between 0.45 and 0.46 and a factor for proportional velocity of between 0.954 and 0.964. Assuming straight line interpolation, these have been calculated as being 0.458 for proportional depth and 0.962 for proportional velocity. This leads us to calculate the depth of flow as

$$0.458 \times 300\text{mm} = 137.4\text{mm}$$

and the proportional velocity as

$$0.962 \times 0.988 = 0.950 \text{ metres/second}$$

Part full pipes by their very nature have a free surface and this can add significant complexity into the calculations as it introduces another variable called the stage. However, for circular pipes, the Colebrook-White formula can be modified to provide a solution. If we assume for both part full and full pipes that friction behaves similarly in each case, then we need to find some parameter or factor for the part full pipe which will be equivalent to the diameter of the full pipe. The hydraulic radius fulfils this criterion and is measured by

$$\text{Hydraulic radius} = \frac{\text{cross sectional area of flow (m}^2\text{)}}{\text{wetted perimeter (m)}}$$

for a pipe flowing full, this becomes

$$\text{Hydraulic radius} = \frac{\text{pipe diameter (m)}}{4}$$

Thus the Colebrook-White formula becomes

$$\frac{1}{\lambda^{0.5}} = -2 \log \left[\frac{k_s}{3.7 \times 4R} + \frac{2.51}{Re\lambda^{0.5}} \right]$$

where

k_s = roughness value (mm)
d = pipe diameter (metres)
R = hydraulic radius
Re = Reynolds number

It should be noted that the discharge from a part full pipe can be higher than that from a full pipe. This is due to the fact that as soon as the pipe ceases to be full, the wetted perimeter reduces far more rapidly than the cross sectional area. Friction is thus reduced and the velocity increases. This higher rate of discharge should be ignored for design purposes because as soon as the pipe becomes full again, the discharge will be reduced back to its original, pipe full rate.

5.4.3 Channels

The choice between a pipe and a channel is usually determined by economic factors, i.e. which option offers the cheapest solution. Channels are a convenient and economic way of conveying large volumes of water and can be used where the site topography is very flat. In hilly terrain and in areas where there are regular significant changes of level, the construction of open channels becomes less cost effective because the alignment of the channel needs to follow the land contours in order to keep velocities low so that the effects of scour and erosion are kept to a minimum.

There are several hydraulic differences between pipes and channels:

- Open channels always have a free surface which means there is less frictional surface area. This condition is only found in pipes in the part full condition
- Channels often have irregular shapes and sections. Pipes have a constant circular shape
- Velocities in channels need to be lower due to the possible effects of erosion and scour. Pipe velocities can be considerably higher
- Water cannot flow uphill in channels. Pipes can surcharge to increase the hydraulic gradient between two points as long as there is a hydraulic gradient upstream and downstream of those points

Shape is an important consideration when specifying channels and although two channels may have the same cross sectional area, the channel boundaries in contact with the water may be significantly different. For example, the two rectangular channels shown in ***Figure 2*** have the same cross sectional areas. The boundary of water in contact with the sides of Channel A however, is greater than that in Channel B.

This boundary, known as the wetted perimeter, is the main source of friction.

Prop. Depth	Prop. Velocity	Prop. Discharge	Prop. Depth	Prop. Velocity	Prop. Discharge
0.01	0.089	0.0002	**0.31**	0.790	0.209
0.02	0.141	0.0007	**0.32**	0.804	0.222
0.03	0.184	0.0016	**0.33**	0.817	0.235
0.04	0.222	0.0030	**0.34**	0.830	0.249
0.05	0.257	0.0048	**0.35**	0.843	0.263
0.06	0.289	0.0071	**0.36**	0.855	0.277
0.07	0.319	0.0098	**0.37**	0.868	0.292
0.08	0.348	0.0130	**0.38**	0.879	0.307
0.09	0.375	0.0167	**0.39**	0.891	0.322
0.10	0.401	0.0209	**0.40**	0.902	0.337
0.11	0.426	0.0255	**0.41**	0.913	0.353
0.12	0.450	0.0306	**0.42**	0.924	0.368
0.13	0.473	0.0361	**0.43**	0.934	0.384
0.14	0.495	0.0421	**0.44**	0.944	0.400
0.15	0.516	0.0486	**0.45**	0.954	0.417
0.16	0.537	0.0555	**0.46**	0.964	0.433
0.17	0.557	0.0630	**0.47**	0.973	0.450
0.18	0.576	0.0710	**0.48**	0.983	0.466
0.19	0.597	0.0788	**0.49**	0.991	0.483
0.20	0.615	0.0876	**0.50**	1.000	0.500
0.21	0.633	0.0966	**0.51**	1.008	0.517
0.22	0.650	0.1061	**0.52**	1.016	0.534
0.23	0.668	0.1160	**0.53**	1.024	0.551
0.24	0.684	0.1262	**0.54**	1.032	0.569
0.25	0.701	0.1371	**0.55**	1.039	0.586
0.26	0.717	0.1480	**0.56**	1.046	0.603
0.27	0.732	0.159	**0.57**	1.053	0.620
0.28	0.747	0.171	**0.58**	1.060	0.637
0.29	0.762	0.183	**0.59**	1.066	0.655
0.30	0.776	0.196	**0.60**	1.072	0.672

Prop. Depth	Prop. Velocity	Prop. Discharge	Prop. Depth	Prop. Velocity	Prop. Discharge
0.61	1.078	0.689	**0.81**	1.140	0.989
0.62	1.084	0.706	**0.82**	1.140	1.000
0.63	1.089	0.723	**0.83**	1.140	1.011
0.64	1.094	0.740	**0.84**	1.139	1.021
0.65	1.099	0.756	**0.85**	1.137	1.030
0.66	1.104	0.773	**0.86**	1.136	1.039
0.67	1.108	0.789	**0.87**	1.134	1.047
0.68	1.124	0.806	**0.88**	1.131	1.054
0.69	1.116	0.822	**0.89**	1.128	1.061
0.70	1.199	0.837	**0.90**	1.124	1.066
0.71	1.123	0.853	**0.91**	1.120	1.070
0.72	1.126	0.868	**0.92**	1.115	1.073
0.73	1.129	0.883	**0.93**	1.109	1.075
0.74	1.131	0.898	**0.94**	1.103	1.076
0.75	1.135	0.912	**0.95**	1.095	1.075
0.76	1.136	0.926	**0.96**	1.086	1.071
0.77	1.137	0.939	**0.97**	1.075	1.066
0.78	1.138	0.952	**0.98**	1.062	1.057
0.79	1.139	0.965	**0.99**	1.044	1.042
0.80	1.140	0.978	**1.00**	1.000	1.000

Table 3 *Factors for part-full pipes*

It follows therefore that the flow rate in Channel A will be lower than in Channel B and this is a useful concept to grasp when selecting the shape of a channel, especially if there are concerns over bank erosion and water velocities are a prime consideration. On the other hand, lined channels should have a shape which gives a small wetted perimeter as velocities are less significant and cost is one of the most important issues. A lined channel can be quite costly to construct, so the least possible amount of concrete should be used in its construction. This lining will occupy the wetted perimeter of the channel plus an allowance for freeboard, so the greater the wetted perimeter, the more concrete will be needed and hence there will be greater costs incurred in construction of the channel. Minimum wetted perimeters are indicated in ***Table 4.***

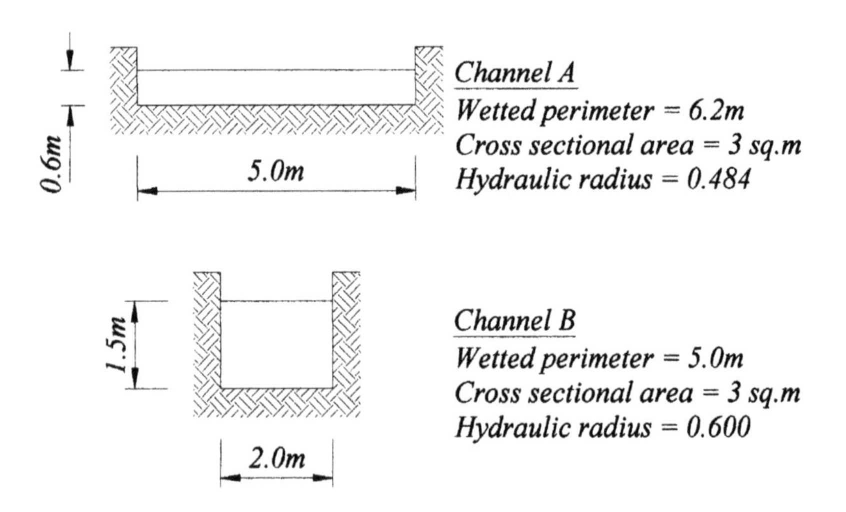

Figure 2 *Comparison of open channels*

Shape of channel	Wetted perimeter
Semi-circular	$4D$
Trapezoidal	$3.463D$
Rectangular	$D^{0.5}$
(where D = depth of flow)	

Table 4 *Minimum wetted perimeters*

To calculate wetted perimeters, the following formulae can be used:

For rectangular channels:

$$p = b + 2d$$

where

p = wetted perimeter (m)
b = bed width (m)
d = depth of flow (m)

and for trapezoidal channels:

$$p = b + 2d(1+y^2)^{0.5}$$

where

p = wetted perimeter (m)
b = bed width (m)
d = depth of flow (m)
y = side slope

To assess the hydraulic shape of a channel, the hydraulic radius is determined from the cross sectional area of the channel and the wetted perimeter. Using the two channels shown in Figure 2, channels A and B have hydraulic radii of 0.484 and 0.6 respectively.

$$\text{Hydraulic radius} = \frac{\text{area (m}^2\text{)}}{\text{wetted perimeter (m)}}$$

In 1926, Fortier and Scobey put forward a paper to the American Society of Civil Engineers which gave recommendations on maximum permissible velocities in unlined channels. A selection of these are set out in ***Table 5***.

In rectangular channels, the maximum discharge occurs when the depth of flow is equal to half the width of the channel. In trapezoidal channels, the maximum discharge occurs when the hydraulic mean depth is equal to half the depth of flow. These are shown graphically in ***Figure 3***.

Channel material	For water transporting...		
	Clear water	Colloidal silts	Non-colloidal silts, sands, gravels and rock fragments
Shale	1.80	1.80	1.50
Pebbles and shingles	1.50	1.70	2.00
Coarse gravel	1.20	1.80	2.00
Stiff clay	1.15	1.50	0.90
Colloidal alluvial silts	1.15	1.50	0.90
Firm loam	0.75	1.10	0.70
Fine gravel	0.75	1.50	1.15
Silty loam	0.60	0.90	0.60
Non-colloidal alluvial silts	0.60	1.10	0.60
Sandy loam	0.55	0.75	0.60
Sand	0.45	0.75	0.45

Table 5 *Suggested maximum velocities for flows in unlined channels*

When using rectangular smooth concrete channels and when the channel width and the depth of flow are equal, the flow can be calculated from the following formulae which are based on the Crimp and Bruges formulae.

$$v = 40w^{2/3}s^{1/2}$$

where

v = velocity of flow in metres/sec and
w = channel width in metres
s = slope

and

$$Q = 40w^{8/3}s^{1/2}$$

where

Q = flow in cubic metres per second

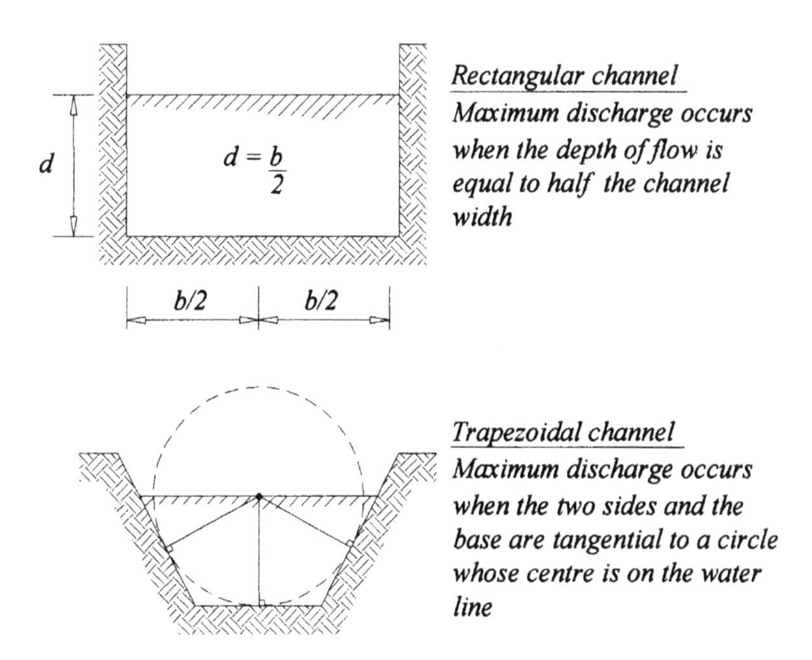

Figure 3 *Maximum discharge in channels*

For the design of open channels, the Chezy formula may be used in conjunction with the Bazin formula. The Chezy formula is:

$$v = C(RI)^{0.5}$$

where

v = velocity of flow in metres/second
C = coefficient (known as the Chezy coefficient)
R = hydraulic mean depth (hydraulic radius) in metres
I = slope gradient (head divided by length of channel)

The Chezy coefficient, C, is not a constant and depends upon the Reynolds number and boundary roughness. A direct comparison between C and λ is given by:

$$C = \left[\frac{8g}{\lambda}\right]^{0.5}$$

In 1869, two Swiss engineers, Ganguillet and Kutter, published a formula for Chezy's C, based upon actual discharge data from studies of channels in Europe and the River Mississippi in North America. The formula employs a coefficient known as Kutter's n and is dependent upon the boundary roughness.

$$C = 0.552\left(\frac{41.6 + 1.811/n + 0.00281/s}{1 + (41.65 + (0.00281/s))n/R^{0.5}}\right)$$

where

s = hydraulic gradient
R = hydraulic radius

For open channel flows, Manning's formula can be used in the following form:

$$v = \frac{0.01\, R^{2/3}\, s^{1/2}}{n}$$

where

v = velocity in metres per second
n = friction coefficient
R = hydraulic radius
s = hydraulic gradient

5.5 Structures of the sewerage system

As with any project, a sewerage system can be broken down into individual components. This section will deal with the principal components likely to be encountered by a drainage engineer and includes the following:

- Pumping stations
- Pumping mains
- Tidal, estuarine and river outfalls
- Manholes
- Gravity sewers
- Inverted syphons
- Flushing chambers
- Storm water overflows

5.5.1 Pumping stations

A pumping station has a specific function to perform and its location is often set by the constraints imposed by the gravity sewers which flow into it.

Pumping stations are frequently sited in low lying areas and for this reason, the effects of flooding of the pumping station should be considered and information should be actively sought as to the highest recorded flood levels in the vicinity and if possible, the pumping station should be sited above the flood level. Where this is not possible, the wet well should be constructed such that it cannot be inundated by flood water, for example by providing upstands or copings around the wet well.

The pumping station should be also located such that in the event of a power failure, any overflow which occurs can be diverted or routed to a watercourse without causing flooding or damage to any property.

The capacity of the pumping station should be such that average, maximum and minimum flows may be accommodated. Variations in flow are generally dependent upon whether the pumping station is for foul or surface water.

Where the pumping station is to be sited on an existing system, records of flows should be obtained from the drainage authority so that the wet wells, pump gear and ancillary equipment can be appropriately and accurately sized.

Any pumping station should have at least two pumps, so it follows that there are two main pumping configurations which can be used:

- Duty/standby
- Duty/assist

Duty/standby operation is, as its name implies, where one pump operates under normal conditions and the other pump operates only when the first pump isn't working e.g. when it has failed. A common variation on this is alternating duty/standby, where the each pump operates alternately. This evens the amount of wear in each pump and means that neither pump stands idle for any length of time.

Duty/assist is when the second pump cuts in to assist the first when flows into the wet well exceed a predetermined level. Again, alternating duty/assist is a common variation. There is usually a third pump required to act as a standby when this configuration is used which will only operate when either or both of the two main pumps have failed.

Fluctuations in pumping rate should be avoided and at least one pump should be sized to accommodate the average inflow into the wet well. This will also avoid frequent cutting in of the pump for short intervals. Standby pumps should be sized so that in the event of failure, the remaining pumps in service can cope with the design peak rate of flow. When failure of the pumps e.g. due to interruptions in power supply is likely to cause significant flooding, or where the consequences of such flooding would be severe, then the installation of standby generators should be considered.

Wet wells should be sized so that at times of minimum flow, the pump runs for reasonable durations and does not remain pumping constantly or for very frequent start/stop intervals. Emergency overflows should be provided at high level from the wet well to a convenient watercourse. These overflows should be screened using either mechanically- or manually-raked screens.

Emergency storage should be provided within the wet well between the top pump start level and the emergency overflow level to allow for occurrences such as pump failure and should be sufficient to provide a nominal volume of 6 hours at Dry Weather Flow (DWF) before the emergency overflow comes into operation or a volume as otherwise agreed with the drainage authority, in the case of public installations; or in the case of private installations, after liaison with the Environment Agency, and drainage board/land drainage authority responsible for the receiving watercourse.

For discharges into a watercourse from either public or private overflows, formal consent from the Environment Agency to discharge is required.

5.5.2 Pump types

There are two principal types of pump:

- Roto-dynamic pumps
- Positive displacement pumps

5.5.2.1 Roto-dynamic pumps

Roto-dynamic pumps rely on rotational movement to provide the pumping action and are also known as continuous flow pumps. They can be classified into three categories as follows:

- Axial flow pumps
- Centrifugal pumps
- Mixed flow pumps

5.5.2.1.1 Axial flow pumps

Axial flow pumps are efficient for lifting large volumes of water at low pressures but due to their relatively high costs, their use is generally reserved for large pumping installations such as for lifting water from a lake to provide irrigation.

These pumps comprise a propeller which is housed in a long and usually vertical tube and which is turned by means of a motor attached to a drive shaft connected to the propeller. The tube acts as the discharge pipe.

5.5.2.1.2 Centrifugal pumps

Centrifugal pumps comprise an impeller which is housed inside a casing. They are also known as radial flow pumps. The impeller rotates at high speed within the casing and vanes on the impeller cause sewage to be thrown outwards into the pump casing where it is collected and delivered to the outlet in the casing.

Centrifugal pumps are probably the most commonly used types of pump and are versatile in application as they can be used for a wide range of situations from low to high heads and with low to high discharge rates. They are particularly suited to applications where a small discharge is required at a high pressure.

5.5.2.1.3 Mixed flow pumps

A mixed flow pump is a compromise of an axial flow pump and a centrifugal pump and combines the advantages of each. The performance characteristics of a mixed flow pump mean that they are more efficient at pumping at higher pressures than axial flow pumps and for large quantities of sewage than centrifugal type pumps.

5.5.2.2 Positive displacement pumps

Positive displacement pumps are simple pumps and can be broadly classified into four types:

- air-lift pumps
- piston pumps
- rotary pumps
- Archimedean screw

5.5.2.2.1 Air lift pumps

An air lift pump forces air into the inlet of a pipe, forcing the contents of the pipe to rise due to the mixture of air and sewage becoming less dense than the sewage itself.

5.5.2.2.2 Piston pumps

A piston pump comprises a piston which moves within a cylinder. As the piston rises, water is drawn into the cylinder through a non-return valve. As the piston falls, the water is displaced through another valve in the cylinder allowing it to rise above the top of the piston. As the piston begins its second upward movement, the water above the piston is forced towards an outlet in the top of the cylinder. Piston pumps are generally used to provide domestic water supplies in developing countries.

5.5.2.2.3 Rotary pumps

A rotary pump contains a set of gears which rotate in opposite directions to each other. As the gears mesh together, the water or sewage becomes trapped between them and is forced towards an outlet.

5.5.2.2.4 Archimedean screw

An Archimedean screw comprises a helical screw which is housed inside a casing. As the screw rotates, water or sewage is drawn into the casing and forced towards the outlet.

5.5.3 Pumping mains

In its simplest form a pumping main will pump sewage from a point A to a point B under pressure. Velocities and sizes of rising mains should be the optimum to prevent sedimentation and to maintain self cleansing velocity. Normally, rising main velocities should be in the range 0.75-1.8 metres/second at the normal pumping rate to offer the most economical design.

The total pumping head includes the static head, friction head and friction losses and velocity head. The static head is measured as the vertical distance between the lowest level in the suction well to the crown level of the rising main at its point of discharge, its highest point, or the highest water level in the chamber at the point of discharge.

The friction head is calculated as the amount of head which would be required to deliver the flow through the main if it were to flow under gravity. Friction losses in bends and fittings can be allowed for by including the equivalent length of straight pipe producing the same amount of friction as the bend or fitting.

Pump friction losses are generally taken as part of the pump's efficiency and are usually considered during pump selection.

Velocity head is generally quite small but should also be considered.

Duplicate mains may be installed if it is considered practical and may be useful under certain circumstances such as phased developments; to accommodate flows in excess of the design flows when those flows could not be conveyed within a single main within the limits of velocity given earlier; or to provide a standby main in the event of the other being taken out of service.

Mains should be suitably anchored within the ground to obviate any displacement due to the pressure of fluid. Anchor blocks or thrust blocks should be adequately designed to resist sliding, overturning and to provide a suitable anchor for the main. Joints above ground should be of the bolted flange type and should be suitably restrained on concrete blocks at regular intervals and at changes of direction. The most commonly used materials for rising main construction are steel, ductile iron and polyethylene. GRP is also used.

Valves should be provided on mains and generally four types of valve are installed. Reflux valves reduce back flow and water hammer and should be located immediately above the pumps on the horizontal section of the main outside the wet well. These are normally located in a separate valve chamber with a cover at ground level.

Air release valves should be provided at each high point in the rising main to allow air to be released from the main when the sewage lying between high points (i.e. in low spots) is driven forward to occupy the space previously occupied by the air when the pumps cut in.

Sluice valves or isolating valves should be located above the reflux valve so that the main can be isolated as required, for example to clear a blockage. Specific lengths of the main itself may also require to be isolated for example either side of a railway crossing, motorway or river.

Washout valves should be provided at low points in the main to allow for the removal and washing out of sludge deposits.

In designing a pumping main there are three principal elements:

- hydraulic design and economics
- surge protection

5.5.3.1 Hydraulic design and economics

In the first instance it is necessary to the design discharge and a suitable pump and pipe diameter combination for it. When the pump first starts, it is required to deliver the design discharge against the static head. After the pump has started, however, additional losses are incurred, known as friction head losses and these vary with the discharge. In order to achieve the design discharge, therefore, the static and friction head losses must be added together and the pump head must equal their sum at the design discharge.

To assess the pump/pipe combinations, the pump characteristics can be superimposed upon the characteristics for the pipeline; that is, the head/discharge curve for both the pump and the pipeline are overlaid upon one another and directly compared and the point of intersection of the two curves gives an indication of a possible pump/pipe combination.

At this stage, it should also be noted that the pump should be running as close to peak efficiency as possible at the design discharge. The pump efficiency curve should therefore also be superimposed upon the pump/pipe curves and the intersection point of the pump/pipe curves should lie as close as possible to the point on the pump efficiency curve where the pump reaches its maximum efficiency.

For this reason, it is likely that several pump/pipe combinations will need to be assessed. The above is a simple explanation of a single pump system. For multiple pump systems, the following formulae can be used to calculate the heads and discharges.

For pumps running in parallel:

the pump head is not factored
design discharge = no. of pumps x design discharge

For pumps running in series:

pump head = no. of pumps x pump head
the design discharge is not factored

5.5.3.2 Surge protection

Surge occurs in pressurised pipelines due to flows becoming unsteady.

This condition can must be considered to avoid damage to the pipeline and can occur for example at pump start-up, emergency pump shutdown or sudden pump failure. To overcome the effects of surge there are two approaches. Both reduce the level of surge to an acceptable level.

- hydraulic damping
- mechanical damping

A simple method of hydraulic damping would be to construct a tower on top of the pipeline and open to the atmosphere. As the water or sewage in the pipeline flows under pressurised steady flow, the tower fills to an equilibrium level, corresponding to the pressure in the pipeline, i.e. on the hydraulic gradient. If a second pump cuts in (for example to deliver a higher discharge), the pressure in the pipeline will rise sharply. Without the tower, this pressure would travel as a wave along the pipeline and could have the potential to cause the pipeline to rupture. With the tower in place, however, the equilibrium level in the tower will rise to accommodate the increase in pressure, thus relieving the surge in the pipeline.

Mechanical damping reduces the surge effects in a pipeline by limiting the rate of valve opening or closing to coincide with and reduce the anticipated rate of increase in pressure.

5.5.4 Tidal, estuarine and river outfalls

There are three factors relating to the sewage to be discharged which should be addressed when considering an outfall and these can be broadly categorised as follows:

- Trade effluent content
- Physical properties
- Chemical properties

Trade effluents, by the very nature of the processes from which they arise, may contain hazardous and highly toxic chemicals. Those with effluents which have notably harmful properties should generally be treated prior to their entry into the public sewers and any trade effluent agreement or licence will stipulate this.

Pollution arising from trade effluent discharges presents the following difficulties:

- Discolouration of water
- Toxic and corrosive substances such as acids, alkalis, organic and inorganic compounds
- Unpleasant but harmless substances which may have bad odours or taste or which may develop the same after treatment by conventional methods
- Suspended organic and inorganic matter
- Oils
- Substances which have bacterial, fungal and algal promoting qualities

Physical properties include, for example, the amount of suspended solids present in the discharge.

Chemical properties are slightly more complex and include the biological and chemical oxygen demands (BOD and COD respectively).

The three factors above have widespread effects on fisheries, natural habitats and on the water which may be used for agricultural (e.g. abstraction) purposes, drinking water (after treatment), or leisure (e.g. swimming). Additionally, on tidal rivers and estuaries, the effects will be much more immediate and apparent.

Pollutants may destroy fish and other life forms due to the depletion of oxygen levels within the water caused by the high oxygen demands of the pollutants or because the pollutants have promoted growths which themselves have high oxygen demands.

Rivers may be used for watering cattle or the water may be abstracted under licence to irrigate or water the surrounding fields. Excessive siltation caused by high volumes of suspended solids in the discharge may cause sludges to form. In hot weather these can ferment and rise to the surface, rendering the water unusable.

Sewage may promote plant, algal and bacterial growths which may inhibit flows in watercourses. On a larger scale the growth may be such that over a long period of time, navigable waters become completely choked and non-navigable.

In some tidal waters, the water does not always run out on each tide and fluctuates between two points. In this case, the effects of any pollutants on the watercourse may be significantly worse.

When considering outfalls where there is a variable water level, as much information as possible should be sought on the level of fluctuation. Tidal records and tide tables are available for most towns and for most rivers and classified watercourses, the drainage authority or Environment Agency will have gauging stations which can provide flow and water levels for specific periods.

Tidal currents will be the most significant factor and will have a direct effect on the course of any sewage once it has been discharged from a long sea outfall, although the prevailing winds may also have a direct but less significant effect on tides and the dispersal of sewage. Current meter and float observations will provide valuable information on the likely course of any sewage discharged from such an outfall.

Additional factors when considering a long sea outfall include littoral drift, local erosion which in the long term may affect local currents, significant surface water run off directly from inland surfaces, river estuaries.

Tides and other factors may combine to determine permitted times of discharge from tidal estuarine and river outfalls. When discharge times are restricted, storage tanks will be required. These should be sized to receive the total design flow during the tidelocked period and should incorporate storm overflows set at a level to provide an acceptable level of dilution.

Storage tanks should be ventilated and for maintenance it should be possible to isolate the tanks to prevent inundation during works such as cleansing or repair or should incorporate self cleansing mechanisms such as automatic flushing devices.

Discharge from tanks should be either automatically operated with a provision for manual override or should be controlled manually by valves or penstocks.

On the discharge pipe at the point of outfall, a flap valve is required to prevent the entry of water into the pipe when the tidal head rises above the hydraulic gradient of the pipe.

The flap valve will allow the pipe to empty down to low tide level and the construction of the outfall must be such that flotation of the structure does not occur when the level of the incoming tide submerges the partially empty pipeline.

5.5.5 Manholes

Manholes should be constructed at every change of direction of a sewer, every change in pipe material, size or shape and at every change in gradient. These three factors are particularly relevant on sewers where man entry is not possible.

The spacing of manholes should be no greater than that which is practicable to allow the use of drain rods or scrapers to clear blockages.

On larger diameter sewers where man entry is possible, it is not essential to have a manhole at every change of alignment, although manholes should be constructed in the locations as described earlier and on straight sections of sewer and at junctions.

When considering the manhole spacing on large diameter sewers, the distance which materials need to be carried for repairs and the ventilation requirements for men working in the sewer should be addressed. Further, if debris or materials are to be subsequently removed from the sewer, this should also be considered.

As a rule of thumb, for sewers up to 900mm diameter, the maximum spacing should be 100m and this is a figure borne out in the current (4th) edition of *Sewers for Adoption* for all newly constructed sewers irrespective of size. For larger diameter sewers, a good manhole spacing guide is 30m for every 300mm diameter of sewer.

Manhole chamber sizes will vary according to their location, depth and pipe size. Set guidelines for manholes for particular applications have been laid down in *Sewers for Adoption* and most drainage authorities keep to these guidelines.

The minimum size chamber that a man can work in efficiently is 1.2m along the line of the sewer and 0.75m across or for circular chambers, 1200mm diameter. The width of the chamber should allow construction of benching of adequate width to accommodate a worker's foot. 225mm is generally accepted as a practical minimum and on sewers of greater than 375mm diameter, one benching should be at least 350mm wide.

The minimum sizes of openings into manhole chambers are generally specified by the drainage authority but as a guide the dimension should be ideally between 600 and 675mm for manholes of normal depth.

Where deep manholes are to be constructed, i.e. over 6m from ground level to invert level, safety is a very significant factor in their design and drainage authorities will have their own safety policies for these manholes.

The engineer should consider that breathing apparatus is likely to be required by those entering the sewer, so there should be adequate provision in the shaft dimensions to provide for this. Any operative entering a deep manhole is also likely to be strapped to a harness so a minimum shaft dimension of 900mm is recommended in one direction, preferably along the length of the sewer.

In deep manholes, the chamber should be a minimum of 2m high to allow for a worker to stand and width x length dimensions should be appropriate to allow two men if necessary to stand in the chamber.

If the manhole is excessively deep, then the shaft may need to be made wider and may need to incorporate landings to allow those entering the sewer to rest.

Manhole materials are predominantly precast concrete and brick, although shallow private manholes are usually plastic.

Manhole covers and frames currently need to comply with *BS EN 124* which gives guidance on the strength classes of frames and covers in particular locations. This publication supersedes the old *BS 497* which some engineers feel gave more stringent testing for strength and consequently it is not uncommon to specify that a cover and frame should comply with *BS EN 124* but should also be British Standard Kitemark certified.

Access into manholes is generally by step irons at convenient spacings for footholds and handholds (usually 300mm centres horizontally and vertically) or by ladder in the case of deep manholes. Current philosophy on the provision of access to deep manholes is not entirely clear, with some sewerage undertakers suggesting that no access ladders or step irons should be provided in the construction on the grounds of safety, ensuring access is only by winch.

There are six main types of manhole, many are variations of the basic and simplest form, a straight through manhole which is a chamber constructed on a straight length of sewer and has no side junctions or branches entering it.

The second type of manhole is a junction manhole. These are constructed at the junction of two or more sewers.

Side entry manholes are common on large diameter sewers and manufacturers can provide standard precast concrete side entry manholes for large diameter pipes. These can be used where it is difficult to achieve direct vertical access into the sewer and can be constructed as far away as the closest convenient location permits, although in practice this distance should be kept as short as possible.

Lampholes are rarely used or constructed in current practice and tended to be used extensively on Victorian sewers. Their purpose was to allow lamps to be lowered into the sewer so that internal inspections could be carried out. They also served to signify changes in direction.

Dual manholes (also known as crossing manholes) are constructed where foul and surface water sewers cross each other or have been laid in common trench. These should be avoided as they are a common reason for cross connection and thus pollution, and may also allow foul sewers to become flooded.

In most dual manholes, the surface water is above the foul and either pipe can be carried through the manhole using for example a ductile iron pipe cast through the manhole walls and encased in concrete. For inspection of the encased pipe, a bolt-down and sealed hatchbox or inspection cover may be fitted.

Backdrop manholes are the final type of manhole that an engineer will encounter. As stated earlier, backdrop manholes can be used to achieve abrupt changes of level and to allow sewers laid at a high level to be connected to a sewer laid at a lower level in the most economic manner.

When constructing a backdrop manhole, the pipe at the higher level will enter the manhole at that higher level, but the end of the pipe will normally have a stopper fitted. Just before this pipe enters the manhole, there will be a vertical or near vertical drop pipe of the same diameter which takes the high level pipe down to the level of the lower sewer. At the base of the drop pipe, there will be a short radius bend before the pipe enters the manhole.

If the backdrop manhole is to be constructed on pipes greater than 450mm diameter, or if the difference in level is less than about 2m, then a ramped backdrop is preferable, although these can be difficult to construct satisfactorily and with adequate compaction of the backfill material.

5.5.6 Gravity sewers

Where site topography permits and excluding open channels and ditches, a gravity drainage system offers the most economical method of draining a site to a given point. A gravity pipeline consists of a length of jointed pipes laid to falls between manholes. The pipes and gradients should be chosen to ensure that the risk of blockage is kept to a minimum and this means that for surface water sewers, a minimum velocity of 1 m/sec should be achieved at pipe full discharge; and for foul sewers, a minimum velocity of 0.75m/sec should be achieved at one third design flow.

5.5.7 Inverted syphons

Any inverted syphon will present a maintenance liability if it is incorrectly designed. The fact that the pipes which make up the syphon are below the hydraulic gradient mean that this type of structure is inherently liable to blockage, particularly if flows are low or intermittent and for these reasons the incorporation of an inverted syphon on a sewerage system should be avoided.

Self cleansing velocity is critical and it should be noted that it is seldom possible for a single pipe syphon to achieve self cleansing velocity even on 1/3 DWF minimum and 4 DWF maximum flows. Once solids have settled in the pipe, it becomes very difficult to prevent further sediment build up.

More care is needed when designing inverted syphons on combined and foul systems than on surface water systems due to the higher suspended solids content of the sewage.

If velocities in a syphon are inadequate, then the incorporation of a means of flushing the syphon through should be provided. This may be achieved by the installation of a penstock or other valve on the upstream side of the syphon.

To flush the syphon through, the penstock is closed to allow flows to back up and is then released. The build up of head on the upstream side should be sufficient to cleanse the syphon when the penstock is opened to allow flows to flush through.

Access for cleaning, unblocking and maintenance should also be considered if incorporating an inverted syphon into a system. The accesses should be such that any individual pipe in the syphon can be isolated for maintenance and the fitting of a bolt-down sealed hatchbox over each pipe in the syphon is generally the best method to achieve this. The access point should be also fitted with sluice valves, extended spindles, risers and tees on the pipeline as necessary to avoid the need for man entry into a deep confined space; and should be sited in a location suitable for tanker access.

Table 6 gives guidance on the velocities required to move different size objects within the pipeline. The minimum desirable velocity in a syphon should be 1m/sec. The critical velocity is that in the uphill portion of the syphon where the larger the object to be pushed up it, the greater the velocity required.

Similarly, the steeper this upstream portion, the higher the velocity needs to be to move the same object.

Pipes in syphons on foul sewers should be sized to carry 1.5DWF. As mentioned previously, single pipe systems seldom reach self cleansing velocity, so the first pipe should be sized on this figure and the second pipe should be sized to carry the remainder. It should be remembered that multiple pipe syphons will remain part full and this can lead to septicity, corrosion and odour problems.

On large diameter sewers, three or more pipes may be required. Again the first pipe should be designed for 1.5DWF; the second pipe should be designed for between 1.5DWF and 3DWF and the third pipe and subsequent pipes sized to take the remainder. In order to split the flows before they enter the individual pipes of the syphon, a system of weirs should be installed so that each weir is set to pass flow over it at the relevant design flow and gradually submerges as the flow rises to pass over the next weir.

The weirs should be designed so that any solids are directed down the relevant pipe in the syphon which has been designed for continuous operation. At the opposite (downstream) end of the syphon, a similar weir arrangement should be constructed to prevent the backflow of sewage into the syphon or into any of the pipes which are not being used at that level of flow.

There are two main factors to be considered when designing a syphon.

The first is the design rate of flow, and the second is the hydraulics of the structure.

Nominal size of stone (mm diameter)	Approximate velocity required (m/s)	
	Horizontal pipes	Vertical pipes
50	0.915	n/a
25	0.610	0.991
12.5	0.458	0.762
6.25	0.305	0.457
Sand grains	0.686*	0.229

** Sand grains lift off the invert at this velocity but at lower velocities, sand ripples form but sand does not travel in suspension*

Table 6 *Desirable velocities in inverted syphons*

The design rate of flow should be determined as mentioned previously (i.e. starting at 1.5DWF for the first pipe), but three additional factors need to be determined:

- the maximum flow through the syphon under storm conditions
- the lowest flow rate in dry weather
- the highest flow rate in dry weather

The operating head of a syphon is equal to the difference in water level between a point just upstream of the entrance manhole and downstream of the outlet. Entry, bend and exit losses in a syphon may be as much as the total losses due to friction and this is especially pertinent when the syphon is comparatively short. The friction losses should be calculated along the length of pipeline. It should be remembered that hydraulically, an inverted syphon is a pipe running full under pressure and if losses are expressed as:

$$\frac{v^2}{2g}$$

where

v = the velocity in metres/second in the first (smallest) pipe
g = acceleration due to gravity

and Manning's formula is used for the losses in friction, then the values of v for each pipe can be calculated. Due to the likelihood of the settlement of suspended solids, Manning's n can be taken as 0.015.

The most common materials used in the construction of inverted syphons are ductile iron, steel, cast iron, precast concrete with concrete surround and PE, although cast iron and precast concrete are now seldom used. If PE is specified, then consideration should be given to maintenance as it is a material which can be easily damaged by operations to clear blockages.

5.5.8 Flushing chambers

Flushing chambers are generally constructed on lengths of sewer where the velocity of flow in the sewer is frequently below self cleansing velocity and where blockages and depositions of silt are likely to occur. However, flushing chambers present particular safety hazards brought about for example, by large quantities of water arriving suddenly at a point downstream in a man-entry sewer or a manhole where someone is working.

Current technology means that mobile jetting crews are able to perform the task of cleansing far more efficiently and safely from the back of a van and using less water than a water tanker dumping large volumes of water in as short a time as possible into a flushing chamber; and for these reasons, flushing chambers are generally nowadays constructed only to cleanse off-line storage tanks.

The chamber should be located conveniently above the pipeline and should ideally be situated at the head of the branch. For manual flushing or jetting operations, where it is necessary to insert equipment into the pipe, flushing chambers should be conventional manholes constructed at minimum depth. For simple flushing operations using a tanker, they need be no more than a lamphole type shaft constructed over the sewer.

If the flushing chamber is to incorporate an automatic flushing system, then the structure needs to be specifically designed to house the equipment and reference should be made to the relevant manufacturer.

To achieve adequate flushing, an appropriate quantity of water should be used and this will be dependent upon the size of the pipe and the length of pipe to be flushed. To ensure that all solids are removed, the initial discharge of flush water must give a velocity in the pipe that is significantly greater than self cleansing velocity of 0.75–1.0m/sec. The effect of the flush may continue in a sewer for a distance of up to 300m but will diminish significantly after about 100m.

As a rough guide, ***Table*** *7* gives the quantities of water necessary to achieve a satisfactory flush for a given pipe diameter. As a rule of thumb for pipes up to 600mm diameter, the volume of water required for flushing should be at least equal to the volume of that sewer running at half full for the length of sewer to be flushed between flushing points.

Pipe diameter (mm)	Volume of flush required	
	Gallons	Litres
225	300–400	1360–1820
300	400–600	1820–2730
375	600–800	2730–3640
450	800–1000	3640–4550

Table *7 Flushing volumes*

5.5.9 Storm water overflows

The purpose of a storm water overflow is to relieve a sewer of excess flows arising during storm conditions. Storm overflows are sited on combined sewers and although the resultant overflow is diluted and screened, it is otherwise untreated (i.e. raw) sewage. The storm overflow therefore should not be sited where the discharge is likely to affect potable water supplies and further, specific consideration should be given to the receiving watercourse when locations downstream are used for boating, bathing or other public recreational activity or amenity; fishing or fish-farming, industrial or agricultural non-potable water supply. All overflows from a sewerage system require consent from the Environment Agency and will require screening of some form to prevent solids from entering the receiving watercourse.

Storm overflows may be required to divert the excess flows to a storage structure, where they are retained and returned to the sewer when the flows in that sewer have subsided. These structures are generally off-line storage chambers.

In the design of a storm overflow, care should be taken to prevent solids from being discharged into the receiving sewer, watercourse or conduit and any method of screening proposed to achieve this should not itself present a maintenance liablility.

Screens should therefore be self-cleansing, being either mechanically raked or designed such that accumulation of detritus such as rags is avoided. The accumulation of debris can easily block the overflow. Spiral separators are now becoming increasingly common and there are various patented designs in use.

Dilution of the discharge is an important factor. Where the receiving watercourse is large, then it has been common practice in the past to allow all flows in excess of six times the dry weather flow to discharge directly into the watercourse. Where the receiving watercourse is smaller, then a much greater dilution is necessary.

The hydraulics of storm water overflows can have significant effects in the sewer local to the overflow, particularly on the upstream side. For example, where an overflow is set at a level above the invert of less than 0.6 times the diameter of the sewer, the draw-down effects of the overflow cause a significant increase in the velocity near the overflow and as the level of flows in the sewer rises, the hydraulic gradient near the overflow steepens and is at its steepest immediately upstream from the point of overflow.

To overcome this, a method of controlling the velocity needs to be introduced into the sewer upstream of the overflow, otherwise significant errors in the estimation of the flows over the overflow will arise.

There are three main types of storm water overflow:

- Syphon overflow

- Oblique weir overflow
- Side weir overflow

Syphon overflows are a simple and cost effective method of relieving excess storm water from a foul sewer. They have often been used to regulate water levels and can be built into the side of a sewer. To ensure self-priming, the inlet to the syphon must be located below the operating level of the sewer.

Priming occurs when the water level in the sewer rises above the crest level of the inside of the pipe which forms the syphon. Water flows over the crest inside the pipe and some of the air trapped within the pipe becomes entrained in the flow over the crest. This results in a drop in pressure which occurs in the pipe and water rises up the inlet of the syphon to balance the pressure, ultimately allowing the syphon to run full.

To break the action of the syphon, air must be admitted to it and this is necessary to prevent the syphon from continued operation after the storm conditions have ceased. To achieve this, air must be admitted either when the water level in the sewer drops below the syphon inlet level, or via an air pipe connected to the syphon pipe and the foul sewer. When the water level falls below a predetermined level, i.e. the end of the air pipe in the sewer, air is introduced into the syphon and the action is broken.

Syphon overflows present a maintenance problem with the accumulation of debris around the vent and methods of overcoming this or an alternative design should be considered when selecting a type of overflow.

An oblique weir is a variation of the side weir overflow. The weir is constructed in a chamber, parallel to the through pipe as per a conventional side weir, but at the upstream end, where flows enter the chamber, the weir is set at an angle to the flow so that on entry, a large proportion of the flows pass immediately over the crest.

One of the advantages of an oblique weir over a side weir is that the oblique angle of the weir has a direct effect on the amount of flows which initially pass over it and this can be set to pass more or less flow, resulting in a smaller, more economical chamber.

Further, there is little loss of either velocity or head and therefore oblique weir overflows are eminently suited to flat areas.

In designing the weir, the oblique section is set at a level slightly higher than is required to compensate for the effects of the increase in radial acceleration as the flows hit the weir. Downstream of the oblique section, however, the weir level should be set so that all flows in excess of the flows designed to continue downstream can freely pass over into the overflow pipe.

Side weirs are possibly the most common form of overflow. The weir is constructed in a chamber, parallel to the through pipe and the crest of the weir is set at a level so that all flows in excess of the flows designed to continue downstream pass over it.

The through pipe should act as a throttle and to achieve this, the downstream pipe should be of a smaller diameter than the incoming pipe, but should be of adequate capacity to carry the design flows. This reduction in diameter regulates the velocity upstream of the overflow and means that the side weir acts more efficiently, allowing more flows to pass over its crest.

To further increase the flow over a side weir and to reduce the length of chamber, constructing the weir on the outside of a curve takes advantage of the increase in radial acceleration caused by the sewage hitting the weir and means that shorter lengths of weir may be constructed.

6

Surface water drainage

6.1 Introduction

Surface water drainage is the practice of intercepting, controlling and discharging surface water run-off from specific catchment areas. In its simplest form it may involve a roof gutter collecting rainwater from a roof and discharging it via an outlet to a rainwater pipe which then discharges it into a pipe laid in the ground and which leads to a ditch.

On a larger scale the drained areas (catchments) may cumulatively amount to hectares and in urban areas will include landscaped areas, roofs, private drives, roads and other areas such as parks. Run-off from landscaped areas must not be drained to an adoptable or existing public sewer even if it is designated as being for surface water.

The principle, however, will remain the same: intercept the run-off and via a system of conduits (sewerage) direct the flows to a suitable outfall.

The following sections describe this process in greater detail.

6.2 Surface water sewage

Unlike foul sewage which contains a range of potential contaminants, in the case of surface water, this is simply the run-off from roofs, paved and other areas caused by a catchment's response to a rainfall event.

When a catchment is modified by a development, the effects can be seen in several ways:

- water quality
- flooding

- water resources
- natural habitat

Whenever rain falls on a catchment, any contaminants will be picked up in the surface water run-off intercepted by the drainage system. These contaminants may include dust, oil, litter and organic matter. When these materials are conveyed to an outfall via a sewer, the levels of concentration of these contaminants is very high and can have significant detrimental effects on a watercourse.

It can lead to a rapid reduction in oxygen levels, thus leading to the suffocation of fish and other aquatic life and a high level of silt and other detritus being deposited on the banks and bed of the receiving watercourse.

Spillages from chemical and oil containers can compound these problems when they are flushed into the surface water drainage system.

These problems are further exacerbated when, as a result primarily of ignorance, used oil, garden chemicals, car washing water and other liquids are disposed of via surface water gullies. Incorrectly plumbed dishwashers, washing machines and sanitary appliances also contribute to poor water quality at the outfall.

Flooding is another effect of surface water run-off. When development of an area takes place, the natural pattern of run-off is changed. Usually, the development will result in an increase in the impermeable area and traditional drainage methods convey this run-off to an outfall as quickly as possible. This results in higher flow rates and can result in flooding downstream of the outfall. Balancing ponds and other measures are required to overcome flooding difficulties and are discussed later in this chapter.

With the development of an area and the resultant increase in impermeable surfaces, there is a resultant reduction in the amount of water finding its way naturally into the ground. In some areas, this can have a considerable effect on the volume of water stored in the ground, reducing groundwater levels and base flows in watercourses.

Natural habitats are also affected by developments. The alteration of the natural pattern of run-off may have a significant effect on the habitat of the receiving watercourse, whether it is a ditch or a river.

The increase in flow rates can cause erosion and can cause silts and other materials to be displaced further downstream, upsetting the natural balance of flora and fauna and habitat.

6.3 Design practice

Current practice dictates that to effectively drain an area, the following criteria need to be identified:

- Area to be drained
- Catchment characteristics
- Consideration of future development
- Rainfall statistics

6.3.1 Area to be drained

In the first instance, a drainage study of the catchment area needs to be carried out. Depending upon the size of the site, this may extend beyond the actual site boundaries. The catchment area will be set by physical features and site topography and should be identified on a plan of a scale larger than 1:2500, and ideally showing contours at one metre intervals. A catchment area may contain natural springs, watercourses, ditches, ponds and rivers as well as other man-made features such as culverts and bridges and these are all physical features which will affect the behaviour of run-off from the site. Full details of these should be recorded, for example normal water levels and preceding weather conditions, physical dimensions, flood water levels and valuable additional information on features such as these is often obtainable from local authorities, drainage boards and the Environment Agency.

6.3.2 Catchment characteristics

Engineering hydrology is concerned with the relationship between rainfall and run-off and is subdivided into two main areas: surface water hydrology and groundwater hydrology. The first of these two subdivisions is further divided into two sections: rural hydrology and urban hydrology.

For the purposes of this book, groundwater hydrology can more or less be ignored, unless designing a soakaway; however, surface water hydrology, or the study of the behaviour of surface water run-off influences the storm frequency and duration that a designer needs to consider when preparing a drainage scheme or strategy.

There are further distinct differences between rural and urban catchments. Although both types of catchment area may contain a range of natural and man-made features, in rural areas the catchment will have a low proportion of impermeable areas and man-made drainage systems, whereas with urban catchments, the opposite is true.

In comparison with a rural catchment, in an urban environment where a high proportion of impermeable surfaces is present, the rate of run-off will be higher, as will the volume of run-off and the peak flow. Times of concentration will be shorter and this will mean that shorter, more intense storms will be of critical frequency and duration.

In considering run-off, the designer should also consider the processes in the hydrological cycle: water evaporates from the ocean into the atmosphere and then condenses and falls as rainfall on a catchment. Within the catchment, there are several circulation routes for rainfall. Firstly, it may be intercepted by vegetation and re-evaporated during transpiration. Secondly, it may soak into the soil and enter the groundwater then flow sub-surface to a river or stream. Alternatively, it may flow overland.

The average ground slope will determine how much surface water is likely to run off the site, but this is not the sole consideration when determining run-off. Soil impermeability is also a necessary consideration. Wet, saturated soils and dry, hard soils, for example, will permit a greater degree of run-off than damp loose soils. Another factor which may influence run-off is vegetation cover and type. Wooded areas allow less run-off than a similar area of grassland on a similar slope.

6.3.3 Consideration of future development

In both urban and rural environments, future development will influence the siting and requirements for attenuation, storage and balancing of flows before discharge to receiving watercourse or sewer.

If the drainage scheme forms part of an infrastructure package of roads, drainage and services which will ultimately serve a large retail park, for example, then calculation of the predicted flows from the developed site will be relatively accurate and easy to ascertain.

If the same scheme is being constructed to serve an isolated housing development in a rural area, it is unlikely that there will be any new developments near the site and so future growth is likely to be minimal.

As a third example, let's say that a new town is to be constructed over a ten year period. Again, population growth and development within the new town are the principal factors to be considered and are ones which can be quantified, but to design and construct the sewerage system based upon the whole of the final developed site will result in an uneconomical and expensive scheme and will mean that the sewers will be greatly oversized for the small flows in the early years.

To design to the other extreme and to cater for only those flows which will be immediately present will result in sewers which will become progressively overloaded as the town develops.

In this example, before design works commence, it will be necessary to plan an overall drainage strategy which will to allow the sewers to be constructed economically and phased in conjunction with the growth of the town so that the situations outlined above can be avoided.

Planning authorities have development strategies which set out and describe the areas within their boundaries which have been earmarked for future development and also describe how those areas should be developed, for example, industrial, residential, commercial, etc.

These publications are generally available for a fee, from the planning authorities and can provide an invaluable source of information when considering a large development or regeneration in an urban environment which may be a designated area of growth.

Consideration should also be given to front gardens in existing environments. In urban areas, there is the possibility than in the future, a lot of these will be paved to provide car parking areas. Although the Highways Act 1980 prevents the discharge of surface water onto public highways from a private surface, this has not always been enforced in urban environments and in the foreseeable future, the trend to convert front gardens to hardstandings is likely to continue.

Thus, areas which should contribute to the private drainage system may contribute to directly to the public sewers via the highway and although ultimately the run-off will end up at the same point, the routing of these flows may mean that times of concentration may alter as a result of changes in time/area diagrams. This problem is exaggerated in areas of good soakage where water draining off gardens which have been converted to hardstanding areas enters the sewerage system.

6.3.4 Rainfall statistics

Rainfall statistics have been collected for far longer periods than run-off statistics with earliest known records in the UK dating back to the 17th century. In the 19th and early 20th centuries, the two most common methods of protecting against floods were the use of the largest recorded historical flood and the use of empirically derived formulae.

Much of the recording of rainfall data in the UK is carried out using gauges to measure and record the total precipitation over a period of twenty-four hours. In 1935, Bilham took this information and produced tables which gave the relationship between rainfall duration and rainfall intensity for given return period storms.

These were based upon data collated over ten years from a group of weather stations over the country and although produced a considerable time ago are, in instances, still referred to by engineers, although generally to provide an approximate indication of the likely mean rainfall. More recently, Road Note 35 contains similar rainfall statistics taken from a broader base.

Rainfall stations are prevalent throughout the UK and there are often additional gauges located at reservoirs and hydro-electric power station sites. Rainfall data is published annually in the UK by the Meteorological Office, but the data is generally expressed in millimetres and not as a rainfall intensity.

On request, the Meteorological Office can, however, supply tables which show average rates of heavy rainfall with frequency and duration for specified locations in the UK. These figures are similar to the Bilham tables and are based upon a computer analysis of selected gauging stations.

Rainfall intensity and frequency vary depending upon the geographical location in the UK. Research has shown that in south-east England, intense rainfall events may be more frequent than in other parts of the country; and in mountainous regions, for example in the north and north-west, the average annual rainfall is much higher, with Scotland's mountain ranges receiving up to five times the UK annual average.

This variation increases significantly with storms of longer duration and return period, but for storms of high frequency and short duration, the regional variation is less marked.

There is, however, a limit to the amount of rainfall likely to fall in a given period of time over a specific catchment. This was discussed in a paper to the Institution of Civil Engineers in 1964 (Wiesner) and is referred to as the probable maximum precipitation.

In the UK, three formulae have historically been used to determine rates of rainfall. Two are referred to as the 'Ministry of Health formulae' and were applied to storms of approximately a 1 year return period.

Expressed in approximate metric units, these first two formulae were:

for $t = 5$ to 20 mins;

$$R = \frac{750}{(t + 10)}$$

(this formula was put forward by a Ministry of Health Committee in 1930 for storms of shorter duration as a modification to the formula below)

for $t = 20$ to 120 mins;

$$R = \frac{1000}{(t + 20)}$$

(this formula was originally suggested in 1906 by Lloyd-Davies and is also known as the 'Birmingham Curve' formula)

The above were based on the formula

$$R = \frac{a}{(t + b)}$$

where

R = rate of rainfall in mm/hour
t = storm duration in minutes
a & b = constants

The Ministry of Health formulae do not take into account the storm frequency but approximate the values given by storms which occur on a once-per-year basis. When accurate local data was not available, this was found to be adequate for most drainage systems. These formulae give a very rough indication of rainfall intensity.

The third formula to have been in regular use was the Bilham formula and was published in 1935. This formula related the frequency of a storm to its duration and intensity and expressed in metric units was:

$$N = 1.25t\,(0.0394r + 0.1)^{-3.55}$$

where

t = storm duration in hours
r = amount of rainfall during time, t
N = number of storms of this intensity likely to occur in a 10 year period

Further formulae based upon work originally published by Norris in 1948 were also used extensively and were incorporated into CP2005: 1968: Sewerage. This publication was superseded by BS 8005: 1987, which although still in current use and circulation, has now been withdrawn from publication by British Standards (BSI) and has been replaced by BS EN 752. The Norris formulae were similar to the Ministry of Health formulae as they were based upon the same equation and have been tabulated to form ***Table 8***.

If either the rational (Lloyd-Davies) or the modified rational method is to be used for design, then a mean rate of rainfall is required. Any of the formulae above will provide approximations of mean rates of rainfall adequate for a preliminary design, but studies and research by the Meteorological Office have since revised and updated the statistics which form the bases of expected mean rates of rainfall and storm profiles and these revisions have increased the mean rates of rainfall by up to 7% or 8% for high frequency storms (less than 2 year return period) and by up to 15% for storms of very low frequency. More accurate data can be obtained from the Meteorological Office which can provide a table giving rainfall data for a specified National Grid Reference. It is always preferable to use such tables and once a collection of these tables has been acquired, the data can be interpolated to suit the site.

Current general practice is for engineers to evaluate rainfall intensity based upon the methodology in the *Wallingford Procedure, Volume 4.*

6.4 Determination of the design storm

For many years, a flat rate of rainfall was used to calculate the rate of run-off from urban catchments. The figures varied from 6mm/hr for large pipes to 50mm/hr for pipes in the immediate vicinity of buildings. For small catchments where the time of concentration is likely to be less than 15 minutes, flat rates of rainfall can still be used if a free and unrestricted outfall from the site is allowed.

Intense rainfall usually occurs during thundery storms which are more common in the summer months, whereas in the winter months, storms tend to be less intense. The difference in these summer/winter storm profiles is not related to the annual average rainfall.

The design storm will comprise two variables – return period (frequency) and duration. The return period is often already set down, particularly if the sewers are to be adopted by a drainage authority; however, the storm duration should be selected so that the occurrence of flooding is reduced to an acceptable level.

Standard annotation for storm frequency and duration is given by:

$$MT\text{–}D$$

where

T = storm frequency (return period,)years; and
D = storm duration (minutes)

Thus, a storm annotated as being M5–60 is a storm with a 1 in 5 year return period and a duration of 60 minutes.

Storm return period (1 in ... years)	Rainfall intensity for t = 5 to 20 mins	Rainfall intensity for t = 20 to 120 mins
0.5	$\frac{580}{(t+10)}$	$\frac{760}{(t+19)}$
1.0	$\frac{660}{(t+8)}$	$\frac{1000}{(t+20)}$
2.0	$\frac{840}{(t+8)}$	$\frac{1200}{(t+18)}$
5.0	$\frac{1220}{(t+10)}$	$\frac{1520}{(t+18)}$
10.0	$\frac{1570}{(t+12)}$	$\frac{2000}{(t+22)}$

***Table 8** Norris formulae for calculating rainfall intensity*

In practice, the most generally used storm frequencies are as follows:

- 1 year return period where the site has average slopes
- 1 or 2 year return period for highway drains
- 2 year return period where the site is flat or where difficulties may arise at the outfall
- 5 year return period where there are properties at risk from flooding
- 5+ year return period for flood protection on rivers depending on location and consequences of flooding (often up to 100 year return period in urban areas)

- 25+ year return period for flow balancing (usually up to 100 year return period but sometimes greater, depending upon location and vulnerability of the site)

In addition to the above, the designer will need to demonstrate that there will be no flooding occurring during a 30 year return period storm if the sewer is to be adopted by a drainage authority. Sewers within city centres may require design to lower frequencies as determined by either or both the drainage and local authority for the area.

Studies have indicated that storms with large return periods rarely cover the whole of the catchment of a sewer. The designer should bear in mind, therefore, that the actual run-off from storms of greater than say a 50 year return period may be less than that of more frequent storms which cover the whole catchment.

The critical storm duration is determined by the time of concentration which can be described as the time it takes for a raindrop landing on the uppermost part of a catchment to reach the lower point under consideration somewhere in the lower part of the catchment.

It has been demonstrated (Lloyd-Davies, 1906) that the storm which will produce the highest run-off from a catchment is the storm which has a duration equal to the time of concentration of that catchment. If the storm duration is greater than the time of concentration, then the run-off will be less because the intensity of the longer storm is less and all the catchment area is contributing to the flow. Run-off in this case is therefore directly proportional to rainfall intensity.

If the storm duration is less, then the rain falling on the upper part of the catchment will reach the lower part of the catchment at the end of the time of concentration and the rainfall in the lower section will by that time have ceased.

The Lloyd-Davies method makes two assumptions:

- that the whole of the catchment contributes to the point under consideration at a time after the start of the rainfall which is equal to the time of concentration.
- that the peak rate of run-off occurs at the same time.

The time of concentration t_c can be expressed as:

$$t_c = \frac{\text{length of sewer}}{\text{full bore velocity}} + t_e$$

where

t_c = time of concentration, expressed in minutes
t_e = time of entry, expressed in minutes
length of sewer is expressed in metres
full bore velocity is expressed in metres per second, based upon the likely pipe diameter and gradient

Using the above formula, the time of concentration is calculated using full bore velocity. In practice, base the time of concentration on a flow velocity of 1m/sec, then undertake iterative calculations using pipe sizes and part full velocities derived from the first calculation. If the formula is used directly as above, then the pipes need to be checked to ensure that the velocity is self cleansing at the relevant proportional depths.

The time of entry (t_e) is the time it takes for rainfall to flow across the catchment into the drainage system and is generally between 2 and 4 minutes. The shorter times of entry are applicable to small and steep catchments, but where the site is exceptionally flat or where there are very large, flat paved areas, a figure of 4 minutes would be acceptable.

Times of entry should not normally be estimated to a greater accuracy than 30 seconds; for areas with very small times of concentration and for storms of normal or average frequency, a change in the time of entry of one minute could give rise to a change of as much as 30mm per hour in the design rate of rainfall, whereas for a longer time of concentration, say t_c = 1 hour, the effect of a change to the time of entry of one minute may result in a change in intensity of less than 0.1mm per hour.

6.5 Evaluation of surface run-off

Once the design storm has been determined, the next step is to calculate run-off. The rate of run-off is affected by physical features, as outlined in the earlier section on catchment characteristics, but there are three main factors which influence the rate of run-off:

- Retention of run-off on the surface (e.g. in depressions as puddles)
- Impermeability and infiltration characteristics (e.g. run-off losses through cracks in pavements or through loose soil)
- The performance of gullies and channels carrying the flows to the sewers

6.5.1 Retention

Depressions in surfaces such as gardens, footways, car parking areas and the like can result in quite a significant quantity of water being held back from entering the sewer. These depressions are present in almost all surfaces.

6.5.2 Impermeability and infiltration characteristics

The impermeability of a catchment plays the most significant part in determining the rate of run-off and is influenced by the following factors:

- Urban catchment wetness index (UCWI) (mm)
- Soil Index
- PIMP

UCWI is related to SAAR (standard average annual rainfall) antecedent wetness condition.

The soil index has values ranging from 0.15 to 0.5 and is based on the *Flood Studies Report*. Values may be obtained from 'The design and analysis of urban storm drainage - the *Wallingford Procedure, Vol. 3*'.

PIMP is the percentage of the paved and impervious areas that discharge to the drainage network.

Many tables have been prepared which give impermeability factors for different surfaces and attempts have been made to distinguish between ultimate and effective impermeability. Ultimate impermeability refers to the difference between rainfall and run-off over time. Effective impermeability refers to the impermeability at the end of the time of concentration.

In a paper to the Institution of Civil Engineers in 1906, Lloyd-Davies wrote that 'the impermeable percentage gradually increases with the duration of rainfall'. Thus, as the surface becomes wetter or more saturated, its capacity to retain run-off or allow run-off to infiltrate into it becomes less. In drainage design it is usual to apply a permeability factor and a range of these factors has been set out in ***Table 9***. The figures given in this table can be used as guidance, although for use in conjunction with the Lloyd-Davies (rational) method and the modified rational method of calculation, figures of 90%–95% and 15% are more probable.

When investigating an existing catchment it is common practice to measure the different roofed, paved and grassed areas and to apply this permeability factor to each area. When considering the run-off from a new housing estate, the same factors can be applied with satisfactory results; although for a preliminary assessment of the same estate, the housing density multiplied by a suitable factor mutliplied by the whole site area will give reasonably accurate results.

Road Note 35 published by the Transport Research Laboratory (TRL) describes that in the majority of urban environments, the total area which contributes to a sewer can be taken as those areas which connect directly to that sewer. This means for example, that a path which drains to a permeable surface should not be included as part of the contributing area. When breaking down a catchment into sub-catchments, this should be taken into account.

In the past, the following formula has been used at the request of drainage and local authorities to calculate the impermeability factor:

$$P = 6.4\sqrt{N}$$

where

P = impermeability %
N = number of dwellings per hectare

Computer technology means that a rapid assessment of drainage requirements can be undertaken at the press of a button, although in order to achieve meaningful results a breakdown of the catchments giving percentage roofed, paved and grassed areas which contribute to the drainage system is required, to which impermeability factors can be applied. The cumulative effects of these areas on the drainage systems can then be modelled for storms of different durations and frequencies to arrive at the critical storm.

6.5.3 Conveyance to sewers

Gullies, channels and drains which convey surface water run-off to the sewer all have hydraulic properties which contribute to how and when the run-off reaches the sewer. If a conduit is blocked for example, then run-off is slowed down and possibly retained on the surface. Similarly, a channel may be a swale or a poorly maintained ditch and this will slow down any run-off thus allowing a proportion to infiltrate into the ground.

6.6 Calculating the rate of run-off

The rate of run-off from a catchment obviously depends upon the amount of rainfall falling on it in the first place, although as described previously, run-off is also affected by several other factors. Once these additional factors have been quantified, however, either of the following methods can be used to determine the amount of rain falling on the catchment and hence the rate of run-off from it:

- Rational method
- Hydrograph method

Type of area	*Run-off factor (%)*
Urban areas	
Heavy industrial e.g. ports, docks	100
General industrial	50–90
Residential	30–60
Parks	10–35
Greenfield sites	
Heavy clay	70
Average soils	50
Light sandy soils	40
Vegetation	40
Steep slopes	100
New residential developments	
10 units/hectare	15–20
20 units/hectare	25–30
30 units/hectare	30–45
50 units/hectare	50–75

Table 9 *Factors of impermeability*

6.6.1 Rational method

Although there are several forms of the rational method, the basic principles have not changed in 150 years. The method was first described by T. J. Mulvaney (Ireland 1850) and later by E. Kuichling (USA 1889) and D. E. Lloyd-Davies (UK 1906).

It is thought that the rational method was first used in Ireland in around 1850 and in the USA from around 1889 and in 1906, this method was demonstrated in th UK by Lloyd Davies in his paper to the Institution of Civil Engineers.

The general expression for this formula is:

$$Q_p = CiA$$

where

Q_p = the peak flow rate in litres/second
C = run-off coefficient which is dimensionless
i = rainfall intensity in mm/hour
A = area drained in hectares

The general expression formula was later altered to give the 'modified' rational formula:

$$Q_p = \frac{C_v C_r iA}{0.36}$$

where

C_v = volumetric run-off coefficient
C_r = routing coefficient
Note: A value of C_v is generally recommended as 1.3 for all systems

and which can be expressed more commonly as

$$Q_p = 2.78\, C_v C_r iA$$

Before the publication of the Wallingford Procedure it was generally assumed that the percentage run-off from an area was roughly equivalent to the percentage of the catchment area covered by impervious surfaces – car parks, roofs etc.

Another way of calculating a run-off coefficient was to stipulate a separate value for different surfaces including lawns, pasture, pavements and roofs and calculate an average run-off coefficient for the whole site.

However, the Wallingford Procedure found that these methods

overestimated the volume of runoff entering the drainage network by up to 30% approximately. If the impermeable area only is used in the modified rational formula then C_v has an average value of 0.75 (it may vary between 0.6 and 0.9). It was further found that the rational method could underestimate flows depending on the shape of the catchment and a routing coefficient C_r of 1.3 was introduced. Therefore C_v and C_r are effectively self cancelling in most cases and there is no difference of engineering significance between the rational method and the modified rational method.

C_r is a constant and the modified rational formula is expressed as:

$$Q_p = 3.61\ C_v iA$$

Note: 2.78 x 1.3 = 3.61

It is important therefore not to use a C_v of 1 with the modified rational method as the peak discharge will be overestimated by 30% as a result of the inclusion of C_r in the above formula. An average value of C_v of 0.75 is acceptable where pervious areas alone are considered. The recommended method of determining C_v is given in the *Wallingford Procedure, Volume 1* and is expressed as:

$$C_v = \frac{\text{PR}}{100}$$

The volumetric run-off coefficient is the proportion of rainfall falling on the catchment which enters the storm drainage system as surface runoff. It is expressed as a fraction of the rainfall falling on the paved and roofed areas and has an average value of 0.75, although a more accurate value can be obtained from the following equations:

$$\text{PR} = 0.829 \times \text{PIMP} + 25.0 \times \text{SOIL} + 0.078 \times \text{UCWI} - 20.7$$

and

$$C_v = \frac{\text{PR}}{\text{PIMP}}$$

where

PR = the urban percentage run-off
PIMP = percentage paved & impermeable area
SOIL = soil index (as defined earlier)
UCWI = urban catchment wetness index (mm)

The value for i for any given storm frequency and duration can be found using the method described in the *Wallingford Procedure, Volume 4* and outlined below.

Firstly, the ratio of M5–60 to M5–2 day rainfall is established. This ratio, Jenkinson's ratio, is referred to as r. Next, the ratio of M5–D to M5–60 min rainfall is determined, where D is the duration. This is read from data plotted using the value of r.

Once duration D has been established, the value of T for MT–D is found from tabulated data which relates the MT–D being considered to the M5–D rainfall.

The value of MT–D derived from the steps above is the point rainfall intensity and is then reduced by applying an areal reduction factor (ARF) to obatin the catchment rainfall depth.

The ARF is a function of duration and area. To obtain the design rainfall intensity, the catchment rainfall depth is found from the following formula:

$$i = \frac{MT\text{–}D}{D}$$

where

i = the design rainfall intensity in mm/hour

There are limitations on the use of the rational method: although it provides a satisfactory tool for the design of pipes up to 600mm diameter and for catchments up to approximately 150 hectares, for catchments and pipe diameters in excess of these figures, alternative methods should be used.

The rational method has an inability to simulate a real storm event which results in inaccuracies outside these parameters and the principal reasons for this are:

- the method assumes when calculating t_f that the pipes are running full bore throughout the time of concentration. This is unlikely to be the case, as most will probably be running partially full
- the predicted rainfall intensity varies continuously down the system. In reality, the intensity from a real storm event will vary throughout duration D and the mean rainfall will be continuous throughout
- the actual rate of rainfall will be constant throughout the storm
- there is no variation in the impermeability of the catchment
- the impermeable area is evenly distributed throughout the catchment

There is a further inaccuracy in the rational method brought about during the determination of the run-off coefficient C_v, although this is improved in the modified rational method formula.

It should also be noted that when the rational method is applied to a length of pipe which has no contributing area, the predicted flows for that pipe are reduced.

This reduction is considerably greater than would normally be expected for the normal attenuation of a flood wave and is brought about by an increase in the time of concentration but no change in the contributing area.

To overcome this problem, there is a concept suggested in *Road Note 35* which suggests the use of an '*effective time of concentration*'. This is a time of concentration which includes times of flow only for those lengths of sewer which have a contributing area; i.e. those pipes without contributing areas are ignored.

The use of the rational method is widespread and since its inception, many modifications have been put forward to make allowance for the inaccuracies brought about by the assumptions upon which the method is based.

As already stated, in using the rational method the impermeable area is assumed to be evenly distributed throughout the catchment. In reality, this is highly unlikely to be a true representation of what actually occurs within the catchment.

In 1933, Ormsby & Hart submitted a paper to the *Journal of Municipal Engineers* suggesting the use of an alternative time–area graph. The purpose of this was to incorporate into the rational method a graph which gave a more accurate representation of the distribution of the impermable/permeable areas within the catchment and to compensate for any inaccuracies brought about by that assumption.

Unfortunately, most of the amended time–area graphs tended to overlook the fact that any inaccuracies arising from use of the original graph were balanced out by the assumption that the impermeability of the catchment remained constant.

This meant that the application of an amended time–area graph to the rational method usually resulted in an over-estimation of the amount of run-off. The hydrograph method amended the time–area diagram further to give more accurate results and is described later in this chapter.

6.6.1.1 Application of the rational method

The procedure starts as one of trial and error, with the preliminary calculations based upon assumed pipe diameters and gradients to give a first estimate of t_c. The calculations may need to be repeated several times until the calculated sewer diameter is equal to the pipe diameter assumed corresponding to the t_c chosen.

When applying these methods, the following steps should be considered and should be read in conjunction with the examples shown in ***Figure 4*** and ***Table 10*** which gives a worked example of a modified rational method calculation sheet.

Step 1

Prepare a plan of the sewer system (often called a network schematic plan) and label each sewer using a decimal reference number. The pipe at the upstream end of the system should have number 1.000 and the sequential system of downstream pipes which gives the largest value of t_c should be sequentially numbered, e.g 1.000, 1.001, 1.002, etc.

Step 2

Calculate the areas contributing to each length of sewer. Make allowance for any areas outside the site which may contribute to the sewer. The contributing areas can be calculated either by measuring each paved and roofed area individually or by taking the whole area and applying an impermeability factor.

Step 3

Determine each pipe length, diameter and gradient. Calculate the full bore velocity for each pipe length and thus the time of flow, t_f, along that pipe. Working sequentially downstream from the first pipe run, add the time of entry (t_e) to obtain t_c, the time of concentration for that pipe. Note that t_c is cumulative and will increase as the calculations progress downstream. Stop the calculations at the first branch, which should be numbered 2.***, go to the top of the run of that branch (i.e. 2.000) and re-commence calculation of the full bore velocities, times of flow and times of concentration sequentially downstream for that branch.

Step 4

At the end of the branch, the cumulative contributing area for that branch should be added to the cumulative area for the branch it connects to. Note that the time of concentration for the whole of branch 2.000 downstream should be less than the time of concentration for branch 1.000 down to the point where it meets branch 2.***.

Step 5

Calculate the rainfall intensity for the time of concentration of the pipe. Multiply the rainfall intensity by the contributing area to give the run-off in litres per second.

Step 6

Compare the calculated run-off with the capacity of the pipe to ensure that the pipe is not overloaded and achieves a self-cleansing velocity; i.e. capacity of the pipe (litres per second) at the specified gradient needs to be greater than the actual run-off and the velocity (proportional and full bore) needs to be greater than 0.75 metres per second.

Step 7

Go back to Step 1 using a value of t_c based upon the previous step calculation of proportional velocity.

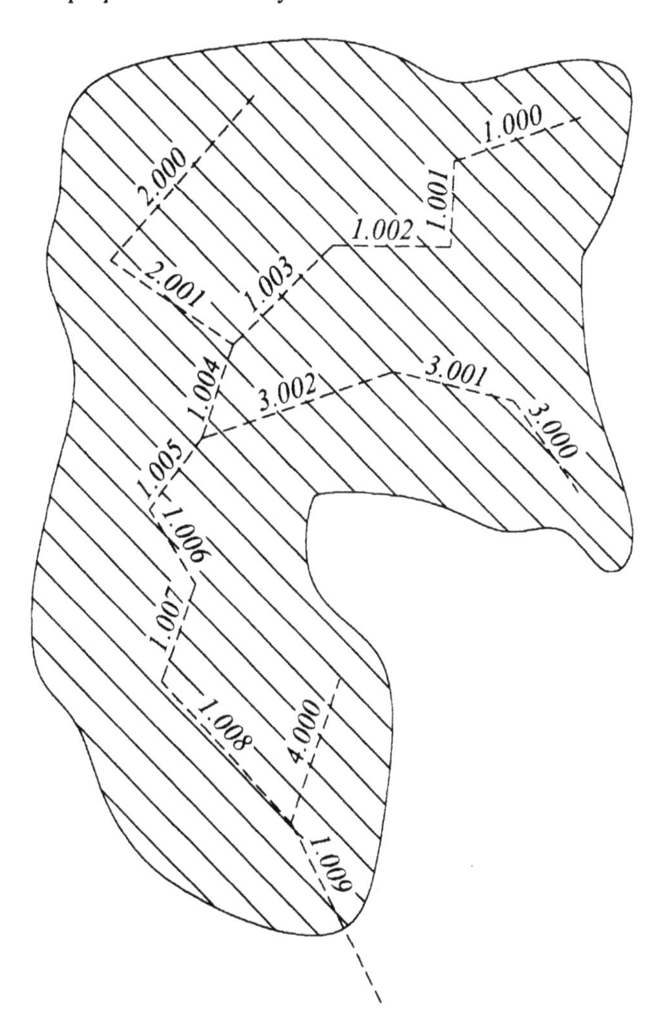

***Figure 4** Typical catchment, pipe layout and pipe labelling*

Pipe roughness value (ks value) = 0.6mm											Peak flow = 3.61Cvia		Cv = 0.75
Pipe label	Pipe length (m)	Pipe gradient (1 in ...)	Pipe dia. (mm)	Pipe full velocity (m/sec)	Pipe capacity (l/sec)	Time of entry (mins)	Time of flow (mins)	Time of conc. (mins)	Rainfall intensity (mm/hr)	Area (ha)	Cum. area (ha)	Peak flow (l/sec)	Comments
1.000	35	125	150	0.896	15.832	4.0	0.65	4.65	58.0	0.075	0.075	11.78	Pipe dia. OK
1.001	30	60	225	1.82	89.6		0.275	4.93	56.8	0.025	0.100	15.38	Pipe dia. OK
1.002	60	50	225	1.853	73.668		0.54	5.47	55.0	0.002	0.102	15.19	Pipe dia. OK
1.003	50	100	225	1.306	51.924		0.64	6.10	50.0	0	0.102	13.81	Pipe dia. OK
2.000	60	80	150	1.123	19.847	4.0	0.89	4.89	56.9	0.125	0.125	19.26	Pipe dia. OK
2.001	42	500	225	0.577	22.923		1.21	6.10	50.0	0.150	0.275	37.23	Upsize the pipe
1.004	25	250	225	0.821	32.637		0.51	6.61	49.1	0.35	0.730	97.04	Upsize the pipe
3.000	27	100	225	1.306	51.924	4.0	0.34	4.34	59.9	0.1	0.100	16.22	Pipe dia. OK
3.001	31	83	225	1.432	56.934		0.36	4.71	57.9	0.015	0.115	18.03	Pipe dia. OK
3.002	45	200	300	1.107	78.217		0.68	5.38	54.6	0.005	0.120	17.74	Pipe dia. OK
1.005	26	200	300	1.107	78.217		0.39	7.00	48.7	0.500	1.350	178.00	Upsize the pipe
1.006	24	83	300	1.722	121.72		0.23	7.23	48.4	0.130	1.480	193.94	Upsize the pipe
1.007	31	125	300	1.403	99.202		0.37	7.60	47.2	0.026	1.506	192.46	Upsize the pipe
1.008	42	100	300	1.571	111.03		0.45	8.05	46.1	0.008	1.514	188.97	Upsize the pipe
4.000	35	100	150	1.003	17.729	4.0	0.58	4.58	58.8	0.005	0.005	0.80	Pipe dia. OK
1.009	48	50	375	2.566	283.45		0.31	8.36	45.2	0.125	1.519	185.89	Pipe dia. OK

Table 10 *Worked example of modified rational method calculation sheet*

6.6.2 TRRL hydrograph method

The TRRL hydrograph method is described fully in *Road Note 35 – a guide for engineers to the design of storm sewer systems.* This was first published in 1963 and has been continuously developed to bring it in line with current standards. The publication also incorporates a section on the rational method and offers a comparison between the two methods of calculating run-off.

The TRRL hydrograph method was developed so that engineers could provide accurate designs for urban storm sewer systems and is generally carried out by computer. The method can be used to determine the sizes required for new sewers and can also model existing sewer networks.

When using this method, in common with other methods, there are several principal items of data which are required:

- Time of entry (in minutes)
- Pipe roughness (k_s value, in millimetres)
- Contributing area (in hectares)
- Pipe length (in metres)
- Pipe gradient
- Storm return period

6.6.2.1 Principles of the hydrograph method

The purpose of the hydrograph method is to be allow prediction of the relationships between rainfall and run-off for any storm frequency or duration. The method has been found to work for a range of conditions and is used extensively by engineers for the design of sewers and prediction of floods.

To describe the method in simplistic terms, observations of rainfall, run-off and time can be plotted as a graph to represent a catchment's response to a rainfall event as shown in ***Figure 5***.

The area under the plotted curve of the graph represents the volume of water being discharged from the catchment and if the base flow (i.e. the flow which is discharged from the catchment in dry weather) is known, then the additional volume of water caused by the event can be determined.

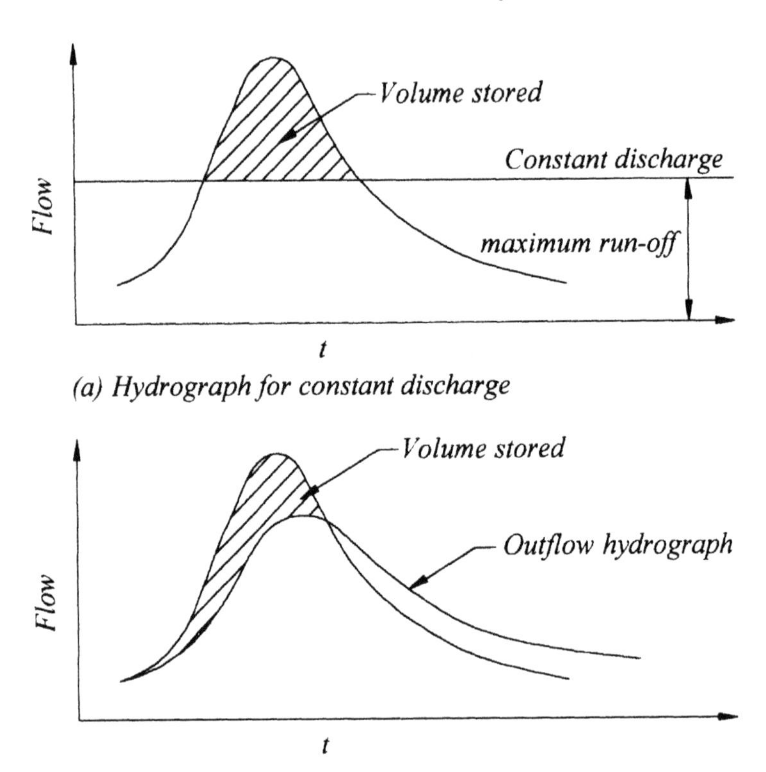

(a) Hydrograph for constant discharge

(b) Hydrograph for varying discharge (e.g. weir)

Figure 5 *Rainfall hydrograph*

If a similar graph is plotted for a point further downstream, it will be found that the resulting curve is longer and flatter, although the volume of water will be the same.

The concept of the unit hydrograph was developed to overcome errors in the rational method and was introduced by Sherman in 1932. Two years earlier, Ormsby and Hart had suggested a revised time–area diagram for the rational method to overcome the inaccuracies arising from its principal assumptions that the distribution of contributing areas within a catchment is uniform and that the rate of rainfall is constant; unit hydrograph theory took the concept of revising the time–area diagram several stages further.

The hydrograph method is based upon three assumptions which lead to the incorrect suggestion that the response of a catchment to a rainfall event is linear.

The first assumption is that at any given point on a system, the base of the hydrograph of the direct run-off from a storm of duration D is constant, irrespective of the volume of run-off. The second is that the ratios of run-off are equal to the ratios of net rainfall intensities but only if the rainfalls are of equal duration. The final assumption is that the hydrograph which represents several events is equal to the sum of each of the individual events it represents.

In applying the unit hydrograph method, the units used are P mm and D hours. The values are usually 10mm and 1 hour for P and D respectively, although any values can be used. The hydrograph is made up of two parts – surface run-off and base flow. Surface run-off responds very quickly to rainfall events as it is formed directly from rainfall as can be deduced from the peak shown in the hydrograph.

Base flow, however, is supplied from groundwater and does not generally respond quickly to rainfall events. The base flow shown in the hydrograph demonstrates the assumption that it remains constant throughout the storm and has a value equal to that which existed prior to the storm.

Just as a hydrograph is made up of surface run-off and base flow, a rainfall event can itself can be broken down into two parts – net rainfall and rainfall losses.

Net rainfall is the rainfall which falls onto the ground and becomes surface run-off (also known as effective rainfall). Rainfall losses make up what is left and can include rainfall that is lost to evaporation or may be rainfall that has infiltrated into the soil and has entered the groundwater.

The relationship between effective rainfall and rainfall losses will therefore affect the volume of surface water run-off. Rainfall losses are closely linked to catchment impermeability discussed earlier.

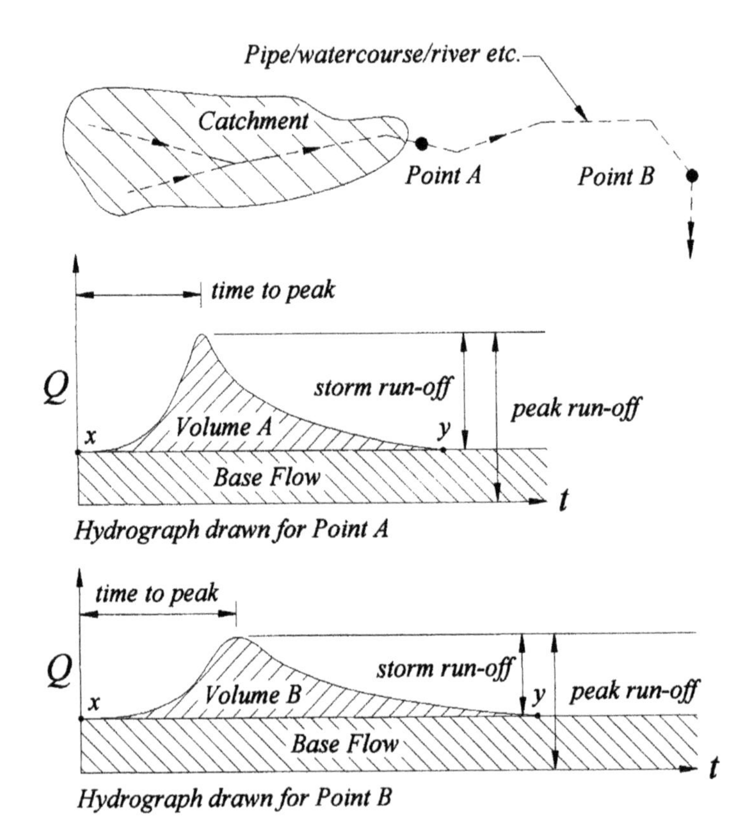

<u>Hydrograph observations:</u>
Time to peak flow for Point B is longer than for Point A.
Peak run-off measured at Point B is lower than at Point A.
Base flows measured at both points are identical.
Volume A = Volume B.
The run-off from a storm passing over the catchment takes time t to wholly pass through a given point e.g. Point A. Time t increases the further downstream it is measured e.g. Point B. Assumption 1 is demonstrated by the line xy which has a constant value for the duration of the storm irrespective of time t.

Figure 6 *Rainfall, run-off and time*

Methods for separating rainfall and base flow have been developed and are generally simple in their application. One such method is demonstrated in ***Figure 6***.

Surface run-off begins at point *x*. The base flow curve is extended to point *A*, the time of peak run-off. Point *B* is determined as the point at which surface run-off stops. A line is then drawn from *x* through *A* to *B*.

The position of *B* is determined by the time measured from the centroid of rainfall to the time of peak. This is known as the lag time. Point *B* is set at a distance of 4 x lag time from the end of the rainfall event.

Once the net rainfall has been determined, the next step is to find the time distribution of the effective rainfall by use of the ϕ index. The ϕ index assumes that rainfall losses are constant throughout the duration of the storm. Losses are measured in millimetres/hour.

A typical rainfall hyetograph is shown in ***Figure 7***. The ϕ index is set so that the P_{net} rainfall is contained above the ϕ index line for the whole of the storm duration. It is highly unlikely that this will ever represent what actually happens in a catchment and two alternatives, an infiltration curve and the concept of percentage run-off, have been used and are also shown in this figure.

6.6.2.2 The TRRL hydrograph method in practice

In use, the engineer first prepares a key plan of the sewer system, as in the rational method. The key plan is then sub-divided into the individual contributing areas for each length of pipe. Individual pipe data and global data (return period, time of entry, etc.) is collated and the engineer can then progress to design. For the design of new sewer systems, the method incorporates four main steps

For modelling existing systems, Step 4 is modified and the purpose of this step becomes to ascertain the status of the pipe, i.e. whether it is surcharged, flowing full or part full, or flooded.

Step 1

A hydrograph is generated for a pipe for which the contributing area has been calculated. This represents the run-off from the contributing area.

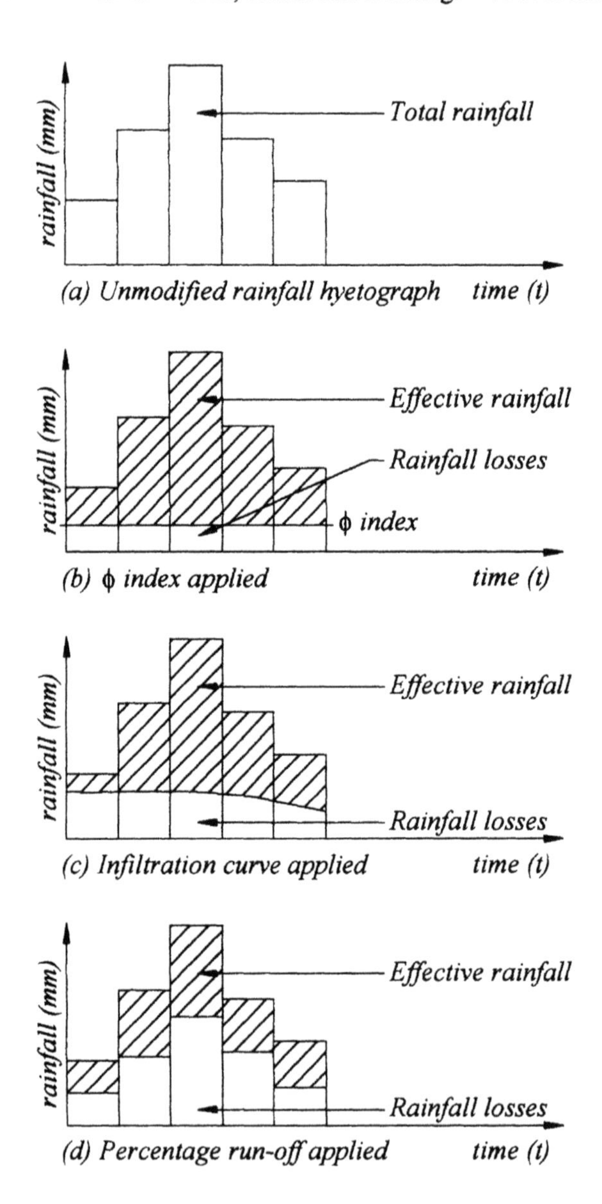

***Figure** 7 Rainfall hyetograph*

Step 2
The hydrograph is then added to the outflow hydrograph from the pipe immediately upstream, if there is one. The two hydrographs are then combined.

Step 3
The combined hydrograph is routed through the pipe under consideration in step 1.

Step 4
The peak flow rate given by the routing of the hydrograph through the pipe is then compared with the flow in that pipe at a proportional depth of 0.85 and if self cleansing velocity is not achieved, then the engineer can upsize that length of pipe.

The effective rainfall depth can be determined from

$$P_{net} = \frac{\text{Volume of Run-off}}{A}$$

where

P_{net} = Effective rainfall depth
A = Catchment area

6.7 Surface water drainage and pollution

The Environment Agency (Scottish Environment Protection Agency in Scotland) is responsible for the protection of controlled waters. Controlled waters include the water contained underground in aquifers as well as the flows in watercourses. There is a high risk of pollution arising from the discharge of surface water and prior to the run-off from urban catchments reaching a watercourse, treatment is required to remove major pollutants.

Run-off from urban areas is not the only potential source of pollution. Temporary works during construction activities can also be responsible for polluting watercourses and before any discharge is made to a watercourse, consent must be obtained from the Environment Agency. Sewage containing untreated foul effluent or trade effluent must not be discharged into a watercourse and is an offence under the Water Resources Act 1991.

One further source of pollution can arise from highway maintenance activities such as the de-icing of roads using salt. Approximately 2 million tonnes of rock salt are used per year for this purpose. The salt by the time it is spread on the highway has been modified and contains anti-caking agents (commonly sodium ferro-cyanide). Other de-icing agents such as urea may also be used.

There are five Acts of Parliament which provide the necessary legislation to control pollution:

- Water Resources Act 1991
- The Salmon and Freshwater Fisheries Act 1975
- Land Drainage Act 1991
- Environmental Protection Act 1990
- Environment Act 1995

In executing any works, precautions should be taken to prevent contaminants from entering rivers, streams, ditches, watercourses and groundwater and these waters should also be secure from siltation and erosion. High risk activities include works near foul sewers, agricultural operations, chemical and industrial plant.

When construction works are taking place, any discharges into watercourses must be controlled. Even activities such as dewatering require consent from the Environment Agency and in some cases lagoons may be required to allow the settlement of solids and silts prior to discharge.

Investigations into past uses of a site will minimise the risk from any operations and can identify if the site has the potential to pollute, for example previous landfill site with hazardous/non-hazardous waste areas, former underground petrol/oil storage chambers on a disused airfield or petrol station. Further, any underground services should be identified and considered at an early stage so that precautions during works may be taken to prevent pollution.

To prevent flooding, any works within sixteen metres of any tidal defence or any works or activities which involve construction or the modification, erection or re-erection of any structure which may interfere with the bed, banks or flood channel of any watercourse or which will take place within eight metres of the bank of any main river require prior consent from the Environment Agency.

Depending upon the type of development, oil separators (or petrol interceptors) may be required and these are proprietary units which can be installed at suitable locations on the piped system.

Sites and developments which normally require oil separation include:

- blocks of garages in excess of 10 units
- large car parks (this can include the cumulative areas of a series of small car parks)
- industrial yards
- oil storage and handling areas
- commercial vehicle and lorry parks
- carriageways

All oil separators should provide at least six minutes, retention and can be of single chamber construction (full retention interceptors) or of the bypass type of interceptor which incorporates an overflow device that allows flows generated by rainfall in excess of 5mm/hour to bypass the separator unit.

Full retention type interceptors treat all the flows that pass through them whereas bypass type interceptors treat only the first flush through the unit as it is this first flush which generally contains he highest concentrations of contaminants. These latter types of interceptor are therefore more suitable for large drained areas where the full retention type interceptor becomes uneconomically large.

In some instances, for example in high risk areas, flow cut-off valves may be required to isolate the interceptor. It should be noted that oil separators, whatever their type, require maintenance and a routine inspection and cleansing programme should be established. Upstream of the unit, silt traps should be installed to prevent excessive amounts of sediment reaching the unit and reducing the unit's capacity which will result in pollutants being allowed to continue through the interceptor.

When considering the drainage design in industrial and chemical process plants, the installation of cut-off valves and bunded areas will provide effective protective measures to combat leaks and accidental spillage.

The disposal of surface water can present a very significant threat to controlled waters in terms of pollution risk. A survey of the quality of river water in Scotland in 1995 (Scottish Environment Agency Report 1996) found that 20% of the poor quality waters were as a result of run-off from urban areas.

The rain falling on urban areas is usually drained to watercourses via pipes, ditches and surface water outfalls. The discharges are often thought of as being clean because the water is not seen to carry pollutants and is often clear. The discharges contain a range of harmful contaminants such as oil, hydrocarbons and toxic metals and may also contain foul sewage from illegal connections and storm overflows.

As a result, the impact on the environment can often be very severe. The design of drainage systems needs to take into account the effects of these pollutants and should minimise the risk of pollution incidents by the incorporation of treatment of the contaminants prior to discharge.

There are a range of techniques available to the engineer to achieve this which are known as sustainable urban drainage systems or SUDS. Local planning authorities are encouraged to include SUDS in local and strategic development plans and developers should also consider SUDS during land purchase negotiations. SUDS techniques can also be used to enhance existing drainage systems prior to their discharge to a watercourse.

For a new development, sustainable urban drainage should be incorporated into a developer's plans at the earliest possible stage as it may have a direct influence on site layout. If the development is seen to have considered SUDS techniques, then this will lie more favourably with the local planning authority when an application for planning approval is made by the developer.

For certain types of development, the local authority may require environmental assessments, in which case it will be the responsibility of the developer to provide an environmental statement. The local authority and Environment Agency can provide guidance on drainage issues arising from such assessments.

Many of the local authorities and sewerage undertakers are committed to the promotion of sustainable urban drainage systems and in addition to the development control process provided by the planning authority, discharges of site drainage may be regulated by the Environment Agency under law covering water pollution.

The regulation of surface water discharges is, however, a discretionary power and the Environment Agency actively encourages the adoption of practices so that smaller discharges need not go through the channels of a formal consent.

A Prohibition Notice issued by the Environment Agency is used to provide control over the larger discharges and can be used by the Agency to require formal consent.

6.8 Sustainable Urban Drainage Systems (SUDS)

Sustainable urban drainage focuses on maintaining and enhancing the quality of the environment receiving a discharge and the people affected by it and has a direct influence on the drainage design.

SUDS techniques involve the incorporation of physical structures in the drainage system designed to receive surface water run-off and include swales, porous pavements, wetlands and ponds. These features should be located as close as possible to where the rain falls and should provide a means of attenuating the high peak normally associated with the shape of the run-off hydrograph. This is known as *'source control'* and may also provide treatment for the run-off using natural processes such as biological degradation, sedimentation, filtration and adsorption.

They can be designed to fit into almost any urban environment or development and the variety of options available with SUDS techniques enables the engineer to consider land use and the needs of the local people when carrying out the design of the drainage system. Ponds, for example, can be designed as a recreational feature or can provide an urban natural wildlife habitat. There are three main subdivisions of sustainable urban drainage techniques but it may not always be feasible to use these techniques to treat all the run-off. These are:

- Source control systems
- Permeable conveyance systems
- End of pipe systems

When assessing the options for drainage design, there are four stages which should be considered for the implementation of a sustainable system.

In the first instance, primary consideration should be given to reducing the amount of water collected in the first place and conveyed downstream. The amount of water determines the sizes of the pipes and drainage system downstream and a reduction will provide the greatest cost benefits. One way of reducing the amount of water is to provide an infiltration system, i.e. a system whereby the water collected can soak into the ground.

The second most important consideration should be to remove the collected water from the site in a manner which reduces the risk of pollution and which promotes further loss of water volume from the system, e.g. by infiltration.

The third consideration is to provide, if necessary, a passive treatment system to improve the quality of the final discharge into the receiving watercourse.

Finally, SUDS cannot be used for a development without considering the long term maintenance. Agreements need to be in place to ensure that the system is adequately maintained. This responsibility currently falls to the private owner, although some local authorities are beginning to adopt these systems.

SUDS techniques provide several advantages over conventional methods for the interception, treatment and discharge of run-off:

- The use of porous pavements allows large flat areas of paving to be constructed without the need for gullies and may eliminate the requirement for pumping
- In certain circumstances, new developments do not need to discharge to existing surface water sewers which may already be at or near full capacity
- The implementation of SUDS may protect existing areas from flooding and may encourage the regeneration of, or development in, urban areas as opposed to development of greenfield sites
- The risk of future flooding is reduced with the implementation of SUDS
- SUDS can be designed to be sympathetic with, enhance and improve

the existing environment

- SUDS can allow natural groundwater resources to be recharged
- SUDS may protect urban watercourses from pollution caused by accidents
- SUDS may remove the entry of pollutants into urban watercourses caused by incorrectly plumbed dishwashers, washing machines and sanitary appliances
- The run-off characteristics from a surface water drainage system incorporating SUDS techniques may be more representative of the run-off characteristics from a natural catchment than the run-off characteristics from a conventional drainage system
- The implementation of SUDS may enhance water quality, natural habitat and biodiversity in urban watercourses
- Widespread adoption of SUDS techniques will see a long-term improvement in water quality in urban rivers
- SUDS cannot be used where the ground water table rises to or is at an impracticably high level

6.8.1 Source control

Source control systems are designed to reduce the amount of surface water run-off from developed sites and new developments. The principal benefits of source control techniques are to:

- recharge underground water supplies
- reduce flood risk
- maintain base dry weather flows in watercourses
- reduce deterioration in water quality

They should be implemented as close to the source as possible and their aim should be to reduce the amount of water discharging to a watercourse or existing sewer as far as is practicable.

Source control systems work best and are most effective when they are distributed throughout small catchments of say 2–3 hectares, where they can intercept small quantities of water close to the point where the run-off occurs.

A good example of this is to construct soakaways and infiltration trenches to directly receive and dispose of the run-off from roofs as this run-off is generally uncontaminated.

There are instances, however, when source control techniques are not appropriate, for example in close proximity to water supply boreholes and major aquifers and on contaminated land.

Where large paved areas or large cumulative paved areas are likely to give rise to significant quantities of run-off using conventional drainage methods, i.e. gullies, manholes and pipes, then the use of porous paving can reduce the need for piped systems and can also remove the contaminants present in the run-off, making it more suitable for discharge.

Further, any filter material under these porous pavements has the potential for storage and can therefore contribute to significantly reducing the size of downstream balancing ponds and tank sewers.

Large paved and roofed areas therefore need not have such a great effect on the hydrology of a catchment as would be experienced using conventional drainage techniques.

With porous pavements as an alternative to conventional pavements, water percolates through the surfacing and then either directly into the subsoil or via a system of porous pipes laid in a filter material to a balancing tank, chamber or outfall.

As stated previously, the filter material can be considered to have a significant storage volume therefore reducing the size of any downstream chamber and this volume can be included by the engineer in storage calculations when the discharge from the site is restricted.

Porous pavements may be proprietary porous or permeable concrete blocks, porous asphalt, gravel, perforated interlocking blockwork or no fines concrete blocks.

With all these systems, if it is necessary to prevent run-off penetrating the formation and entering the groundwater, the formation can be overlaid by an impermeable membrane and the run-off collected by a network of porous pipes which discharge at a predetermined point.

The time of entry of rainwater into these systems is also increased and this means that the resulting run-off hydrograph will be a lot flatter than that from a conventional drainage system. Research has shown that the removal of pollutants by these systems has been considerable and that some are retained in the pavement material, particularly in the case of porous asphalt. Removal rates for trace metals and organic matter are high and the removal of approximately 80% sediment, 80% nitrogen and 60% of phosphorous contents has been recorded.

A further technique used in source control is the construction of an infiltration trench. This is a shallow trench which has been backfilled with stone. Surface water run-off directed into the trench gradually percolates through the subsoil and back into the groundwater system.

Soil characteristics play an important role in the performance of an infiltration trench, with high permeability rates being favourable. Depth to the water table is also an important factor to be considered.

Natural pollutant removal mechanisms occur within and beneath the trench, with microbial action occurring in the subsoil and filtration occurring within the backfill material. The performance of the trench may be enhanced by the incorporation upstream of pre-treatment to remove solids such as a gully, sump or grassed filter strip. The installation of infiltration trenches has been proved to reduce the levels of trace metals and organic matter, solids, phosphorus, nitrogen and coliforms.

An infiltration basin is a variation on an infiltration trench and is a shallow surface depression where surface water run-off is allowed to accumulate and gradually infiltrate into the ground. It may be any shape: long, short or irregular but must retain the run-off it receives. If necessary, an emergency overflow may be provided. As with infiltration trenches, the primary considerations affecting the performance of the basin include soil characteristics and the depth of the groundwater table. Natural pollutant removal mechanisms are much the same as those for infiltration trenches.

6.8.2 Permeable conveyance systems

Permeable conveyance systems slow the velocity of run-off to allow soakage into the ground and to allow sediment settlement which leads to the filtration of solids. There are two principal types of conveyance:

- Swales
- Filter (French) drains

Swales are shallow, wide, grassed depressions and can be used to link storage ponds and wetlands or can form a network and can be used to provide temporary storage prior to discharge.

In addition to performing the environmentally friendly task of removing pollutants, the inclusion of swales in a highway scheme can also be environmentally friendly towards amphibians which are often trapped in road gullies and are only removed when the highway authority's 'gully sucker' pays its regular maintenance visit.

A swale could be described as an open ended infiltration basin which allows surface water run-off to flow overland to storage or to a discharge point.

For optimum performance, swales should have side slopes no steeper than say 1:5 (ideally no steeper than 1:7) and should have longitudinal gradients that are as flat as practicable. Grass length should be kept to around 150mm. These criteria ensure that flow velocities are kept to a minimum thus allowing better removal of pollutants and reducing the risk of erosion.

Swales should be designed to be dry, but as a rainfall event passes over the catchment, run-off gradually enters the swale by natural overland flow. The grassed surfaces reduce flow velocities and filter out sediment and other residues such as oily deposits and organic matter. These are retained and broken down in the top layer of soil and vegetation.

They can often be accommodated in roadside verges where run-off from the carriageway flows over the edge of the road and directly into the swale. In road construction terms, this means that kerbs with upstands are not required and thus brings about a cost saving.

Filter drains, also known as French drains, have been used successfully for many years in highway schemes, industrial and residential developments. They comprise a geotextile lined trench backfilled with a gravel or stone filter material and run-off enters the drain either directly or via a pipe.

Run-off is conveyed slowly through the filter media towards a discharge point such as a watercourse, allowing the filtration of solids and retention of other deposits such as oils and organic matter. Dispersal of water out of the drain, principally through the sides, may mean that an outfall is not required, but this is determined by the soil characteristics and the depth of the groundwater table. Over time, bacterial action breaks down the retained oily residues and organic matter.

6.8.3 Passive treatment systems

Passive treatment systems are 'end of pipe' systems that can be incorporated into landscaped areas and should be used where other sustainable drainage systems cannot be incorporated into the design.

They are described as 'end of pipe' systems because they are sited at the point of outfall but prior to discharge, for example into a watercourse.

Passive treatment systems use natural processes such as biodegradation to remove pollutants from the run-off. On large sites these systems often require the collection and storage of water in ponds, retention basins or wetlands specifically designed to treat the pollutants.

In these instances, the ponds can promote the occurrence of natural pollutant-removing processes and can also provide a natural habitat for wildlife and provide the local community with amenity features.

A wetland is a type of retention pond which incorporates selectively planted shallow areas. These shallow areas allow for a high degree of pollutant filtration by plant, bacterial and algal action but are at risk from siltation if sumps are not incorporated upstream to remove sediment.

Existing natural wetland areas are not suitable for surface water treatment and the introduction of polluted surface water into the ecosystem will have significant and far reaching detrimental effects.

It is not generally accepted practice to direct surface water to an existing wetland; run-off should therefore be directed to wetlands designed specifically to treat surface water. If entirely necessary, with careful ecological consideration and assessment and the incorporation of selective planting design by specialist consultants, an existing wetland can be modified to treat surface water.

Retention ponds are ponds designed to contain water at all times. They can be visually attractive and can provide a useful public amenity, although to be successful as such, the catchment area should be no less than five hectares and there should be a regular base flow into the pond.

To be effective, a retention pond will have a retention time of at least 20 days. This will allow biological degradation of pollutants and can improve the removal of elevated nutrient concentrations which can result in algal blooms, trace metals, organic matter and coliforms. To prevent the ponds from becoming excessively silted, sumps should be provided upstream to collect sediment and downstream to prevent blockage of the outfall.

Some authorities consider that the safety hazards presented by retention ponds are too great and these hazards need to be considered in design. A suitable alternative to a retention pond is a dry ponds or detention basin.

These are designed to remain dry except during or after storm conditions. Surface water run-off is retained in the basin for a few hours after the storm has passed and is allowed to discharge into the receiving watercourse or sewer at a controlled rate. The detention period can be designed to contain the 'first flush' and to allow all other flows to by-pass the basin.

A detention period of twenty four hours will allow for removal of a modest proportion of the pollutants, but the principal feature of a detention basin is its ability to remove solids: solids removal capability is quite high, but its ability to remove trace metals and nutrients is relatively low, although this can be improved by the incorporation of a small wetland or permanent pond into the basin.

6.9 Soakaways

Where it is impractical to discharge the surface water run-off into surface water sewers, for example, an outfall is several miles away or there is no public sewer in the vicinity of the development, if ground conditions permit, soakaways are a viable and practicable solution to the disposal of surface water run-off.

6.9.1 Design considerations

Before considering a soakaway scheme, the engineer should note that it is an offence to allow pollutants to enter the groundwater. Aquifer protection policies have been brought about by the Environment Agency to protect the groundwater supplies from pollution and the powers of the Environment Agency have been further enhanced by the Environment Act which confers additional powers and makes the pollution of groundwater by discharge, spillage or soakage an offence.

Surface water discharges to soakaway systems are subject to control under the Water Pollution Regulations and where List I or List II substances are present, the Groundwater Regulations 1998.

Soakaways should therefore be constructed so that any pollution does not enter the groundwater system and thus aquifers; and should be constructed if necessary after consultation with the Environment Agency. Soakaway schemes should also be constructed so that any pollutants carried in the surface water run-off are removed prior to surface water disposal into the ground.

Later sections in this chapter deal with source control and sewage quality and assist the engineer in preventing the occurrence of groundwater pollution incidents.

The design of any soakaway should take into account as a minimum the following:

- Position of the groundwater table
- Ground permeability and soil characteristics
- Rate of water entry into the soakaway

The groundwater table may vary seasonally, it may be stable or it may rise permanently or locally. In order to function, soakaways must discharge into ground that is above the water table; further, soakaways that discharge directly into the groundwater or where there is very little difference in level between the soakaway and the groundwater table can introduce untreated pollutants into the water supply.

Ground permeability, or the ability of the ground to absorb and disperse surface water run-off is a major factor in soakaway design and has a direct and significant influence on soakaway sizing. Permeability of the ground may increase or decrease with time and in response to the infiltration of surface water run-off into it.

Permeability of the ground also determines the length of time it takes for the soakaway to be ready to receive the run-off from subsequent storms. Soil types and their suitability for soakaways are shown in ***Table 11***.

Soil classification	Suitability for soakaways
Sandstones	Yes, if fissured but groundwater may become perched. Possibly prone to erosion.
Chalks	Yes, but lower and softer grades of chalk are unsuitable (e.g. 'putty' chalk). Consider also the proximity of foundations and adjacent buildings or structures
Clays	No. Particularly London Clay, Weald Clay and Gault Clay and clayey/flinty head deposits
River and beach gravels	Yes
Peat, loam	No

Table 11 *Suitability of soils for soakaways*

The rate of water entry into the soakaway is determined by the area drained and the rate of rainfall. Generally, it is assumed that the total area contributes instantaneously to the soakaway, i.e. no allowance is made for time of entry or time of flow. Design rates of rainfall can be determined from tables such as those published in the *Wallingford Procedure*.

There are several key factors which need to be considered and which are of direct relevance to the engineer. These factors address principally the problems associated with water entry into the ground and can be broadly classified as follows:

- Stability of the ground
- Changes in ground permeability
- The proximity of the soakaway to any structure
- The effects of the entry of grits and silts into the soakaway
- Erosion of the strata surrounding the soakaway

The stability and structural integrity of the ground around a soakaway will vary according to the soil classification and soil type with some soil types being more susceptible to the effects of the entry of water than others.

Over a period of time, the water discharging from the soakaway may cause widening of the fissures into which it permeates, for example, in chalk. The opening of the fissure matrix means that fines may be washed through the soil away from the soakaway and over further time, this can lead to the formation of voids, usually around the base of the soakaway.

Partial collapse of the soil above the void eventually occurs and the cycle continues until the remaining soil between the void below and the surface above collapses through its own weakness or by loadings imposed at ground level.

Understandably, if this situation can occur in an area local to the soakaway, it follows that cumulatively over a period of time, the effects of water infiltration into the soil may reduce the bearing capacity and permissible yield stresses in the soil on a wider scale.

This becomes more apparent if a structure is located on the downstream side of a soakaway. Where hydraulic gradients are vertical as in highly permeable and fissured soils, the effects of fines migration is likely to remain local to the soakaway, but where groundwater flows are more lateral, then the likely widespread effects should be considered.

Chalks frequently offer excellent soakage potential and are often eminently suitable for the installation of soakaways. They are, however, particularly susceptible to instability: groundwater percolates into the chalk, the chalk matrix opens and softens, fines are washed through and a void is formed.

This can occur over relatively short periods of time particularly if the source of water is concentrated (e.g. a soakaway) or it may take several hundred years if allowed to occur naturally. The voids formed are known as solution features (also dissolution pipes, swallow holes or cavities) and are generally metastable features, that is they are stable until some event occurs which pushes them out of equilibrium and causes a collapse. In the case of a soakaway, this event can be a storm or a wheel load.

Ground permeability may change once surface water from a soakaway starts to penetrate. Site soakage tests may reveal for instance that sandy materials such as Folkestone and Thanet beds will provide good soakage rates suitable for the construction of soakaways, but these soils often silt up very quickly and are often prone to the collapse type settlement as described earlier.

Soakaways inherently present higher maintenance requirements than positive drainage systems with an outfall pipe. The principal reason for this is the entry of silts and grits into the soakaway chamber which, over time, builds up and reduces the effectiveness of the soakaway.

Surface run-off naturally transports sediment and other detritus, especially during the first flush after a prolonged dry period. Once this has entered the soakaway, the effects can be two-fold.

In the first instance, the transported silts can migrate through the drainage media surrounding the soakaway and can cause the soakaway to be come blocked.

Secondly, the build up of deposits within the soakaway chamber reduces the available surface area through which the run-off can disperse, thus reducing the capacity of the soakaway.

Silt traps, trapped gullies and catchpits should therefore be specified upstream of the soakaway if these difficulties and maintenance liabilities are to be avoided. Soakaways should be cleaned on a regular maintenance cycle rather than left until the problems of siltation manifest themselves in the form of surface ponding.

Soakaway cleaning may present a difficulty itself, particularly when the soakaways are of the filled trench type and in this instance, a design life of, say, fifteen to twenty years should be assumed, after which time the soakaways should be dug out and replaced.

There is one housing estate near Brighton, East Sussex which was constructed in the 1960s and where all the surface water drainage was directed to soakaways. The pitch fibre pipes which made up the sewer network have since been replaced but the soakaways continue to perform reliably and have not undergone any maintenance since they were first constructed over thirty years ago.

The choice of soakaway type is generally determined by what can be practicably constructed to serve the area drained for the specified storm and there are four types of soakaway available to the engineer:

- Trench soakaway
- Domestic rubble filled soakaway
- Conventional soakaway
- Borehole soakaway

The first two types can be dealt with as one topic, as the domestic rubble filled soakaway is little more than a scaled down version of the trench soakaway.

These soakaways are simple excavations filled with clean rubble or crushed stone and can be used where the ground has good soakage potential. They are relatively cheap and easy to construct and are most commonly used where the areas drained are small. They are not generally adopted by drainage authorities and are usually located in areas that are in private ownership.

When sizing filled trench soakaways, consideration should be given to the void ratio of the proposed fill material as this will determine the volume required within the soakaway. As a guide, a void ratio of 30% achieved using clean rubble will mean that the actual volume of excavation below the invert of the incoming pipe will need to be:

$$\frac{100}{30} = 3.333 \text{ times greater than an unfilled soakaway}$$

Where larger areas are to be drained, or the soakage potential of the ground is only fair, then a conventional soakaway may be constructed.

A conventional soakaway may be an unfilled lined excavation such as perforated precast concrete sections or open jointed honeycomb brickwork. The former are generally preferred by highway authorities when the soakaway is to be offered to the highway authority for adoption.

A borehole soakaway is essentially a conventional soakaway situated over borehole with a perforated liner taken down to suitable strata. The conventional section of the soakaway is generally a precast concrete ring which may or may not be perforated.

6.9.2 Principles of soakaway design

There are several methods of testing for soakage which can be applied to determine the sizes of soakaways, but when testing is not carried out, and domestic soakaways are designed based upon the knowledge of the surrounding ground conditions and local knowledge, a flat rate of rainfall is assumed for the area drained. This method should only be used when sizing domestic soakaways. The principles of soakaway design are based upon the following parameters:

- Flat rate of rainfall for use with domestic soakaways
- Design storm for filled trench soakways
- Soakage test results

Domestic soakaways are generally sized using a flat rate of rainfall of 12mm/hour over the entire area drained. This figure can also be applied to conventional soakaways used in domestic applications.

In domestic and non-adopted applications, the design storm is often taken as a return period and the volume of storage in the soakaway is calculated for a range of durations and is determined by the soakage rate out of the soakaway. For soakaways in adopted applications e.g. highway soakaways, the design storm will usually be specified by the adopting authority.

There are three main types of test for soakaways. The first is a standard test set down in BRE365 and involves the excavation of a pit in the base of the proposed soakaway. It should be noted that this test is not suitable for borehole soakaways. The pit is then filled with water and the time it takes to empty to half full is recorded. This test is carried out several times and gives an indication of the soakage out of the soakaway.

The second type of test is a falling head test. In this test, a borehole is sunk to a specific depth. A perforated liner is situated within the borehole casing and the casing is then withdrawn to expose a given surface area of perforated liner within the borehole. The borehole is then filled with water and the time it takes to fall to the level of the top of the perforated liner is recorded at regular intervals. The results are then plotted on a graph and a soakage rate is obtained for the borehole.

The third type of test is a variation on the falling head test. Again, it is used to determine the soakage rate out of a borehole soakaway. The procedure is exactly the same as for the falling head test, but instead the water level is maintained and the rate at which water is added to maintain a constant head is observed at regular time intervals and plotted as a graph to determine the soakage.

There are several readily available publications which give further guidance on the design of soakaways

- *BRE Digest 365* is a current document produced by the Building Research Establishment which gives guidance on the design of filled trench and precast concrete type soakaways
- *CIRIA Report 156 – Infiltration Drainage* also provides guidance on these types of soakaways
- Soakaway design is also considered in *BS8301* the British Standard Code of Practice for Building Drainage
- For borehole soakaways, there is a technical memorandum published by Kent County Council, *The Design and Installation of Soakaways in Kent (Technical Memorandum M86/1)I*; although this publication is primarily aimed at soakaways likely to be adopted by Kent County Council in the course of carrying out its functions as an authority, the guide contains information relevant for many borehole soakaway applications.

6.10 Bibliography

Design and Analysis of Urban Storm Drainage, the Wallingford Procedure volumes 1–4, published by Hydraulics Research Limited, Wallingford

The Flood Studies Report, published by the Institute of Hydrology, Wallingford

7

Highway Drainage

7.1 Introduction

Highway drainage, although essentially designed on the same principles as surface water drainage for urban catchments, differs in the nature of the run-off and the fact that all drainage is shed from the road surface by the camber or cross-fall of the road. Highway drainage is the responsibility of the highway authority and not the sewerage undertaker, unless the pipe conveys water that is predominantly from roofs and other off-highway paved areas.

Highway drainage conveys surface water run-off only and should be designed so that the carriageway is drained effectively – surface ponding can represent a significant hazard to road users and it is therefore necessary to provide adequate means of removal of surface water.

The surface of a highway should be regarded as wholly impermeable, that is, all paved areas of the highway contribute to the drainage system with 100% run-off. For large and relatively flat unpaved areas such as verges which drain to the highway, run-off can be reduced to 15 to 25%. Embankments without drainage at the toe of the slope should be considered to have up to 100% run-off, depending upon the steepness of the slope.

As with surface water drainage design, it is necessary to determine the amount of run-off by consideration of a design storm. For highways, it is usual to consider a frequency of either one or two year return period. For approximation and for reasonably accurate calculations of run-off from small lengths of highway, mean rates of rainfall can be taken from a table such as that published in *Road Note 35* or graphs based upon a formula such as Bilham's formula.

Drainage design for major highway schemes such as infrastructure, motorways and trunk roads should be undertaken using a method such as the modified rational method, the TRRL hydrograph method or a computer simulation method.

Times of entry will generally be similar to those for urban catchments, but it is generally recommended that the time of entry for unpaved areas contributing to the highway drainage should be taken as 60 minutes. This is due to the fact that the peak flows in sewers resulting from storms of summer profile are not significantly affected by the run-off from these unpaved areas, although where the unpaved areas are embankments and the road is in cutting, this long time of entry is not likely to be appropriate and a figure of 2–6 minutes would be more realistic.

Highway drains are the responsibility of the highway authority (or the Highways Agency in the case of trunk roads) and are not the responsibility of or maintained by a sewerage undertaker until they convey surface water run-off from a private area (e.g. a roof, drive or private road).

There are four principal methods of draining a carriageway:

- conventional kerbs and gullies
- construction of porous pavement
- run-off to channels
- over the edge drainage

Whichever method is employed, the carriageway needs to be drained effectively so as to cause the least amount of inconvenience and hazard to road users, whether vehicular, pedestrian or cyclist. With all systems, the drainage must be adequate to drain the carriageway without excessive ponding on the surface. The level of acceptable ponding will depend upon the category of road and its location. For example, in an urban high street environment adjacent to a busy bus stop the acceptable level of ponding will be less than that for a rural carriageway where there are few pedestrians. Notwithstanding this, the designer should be aware that ponding on the carriageway may cause vehicles to aquaplane and thus lead to a high accident rate.

It should also be noted that good drainage prevents softening of the ground at the edge of the carriageway and thus prevents edge and foundation failure.

7.2 Conventional kerbs and gullies

Conventional kerbs and gullies are probably the most common form of carriageway drainage and are used throughout the UK on estate and other roads. With this type of system, it is essential that the spacing of gullies is considered so that any water flowing along the channels is intercepted and removed from the surface without ponding. Where gullies are used, kerbs provide a positive restraint for any surface water flowing in the channel and it is this depth of water which needs to be kept to a minimum.

In the past, considerable use was made of the rule of thumb that one gully could drain $200m^2$ of paved area. Research has shown that this is not entirely accurate, although for most residential and industrial estate roads this rule continues to be used by many engineers without undue concern and is frequently specified in estate road design manuals.

In practice, on urban and rural carriageways other than estate roads, gully spacing is a more exact science and a procedure has already been established for determining the required spacings of gullies located in carriageways for various carriageway crossfalls and longitudinal gradients.

From experiments carried out by Hydraulics Research Station, the following points were noted:

- when the width of flooding at the channel is equal to or less than the width of the gully grating, the gully grating is highly efficient
- the gully efficiency drops as the width of flooding, flow velocity and the longitudinal gradient increase
- the discharge from the gully continues to increase with increased width of flooding due to the increased depth of ponding
- an increase in the length of the gully parallel to the channel is less effective than an increase of the width of the gully perpendicular to the channel

When designing gully spacings, there are five main points which will determine the results:

- the location, geographically of the road. This factor determines the amount and intensity of rainfall falling on the catchment

- the length of the carriageway if the whole carriageway is of uniform crossfall and longitudinal gradient, or the length of carriageway divided into sections of uniform crossfall and longitudinal gradient
- the type of road surface (concrete or bituminous materials). This factor determines the roughness of the surface and influences the rate of run-off. It is equivalent to Manning's n-value
- the required drainage design parameters to include the maximum flood width at the channel and the storm frequency
- traffic conditions. This factor determines the class of gully required

The conventional method of draining a highway is by shaping the road surface so that it sheds water to one or both sides of the carriageway where it can be collected by a channel or a series of gullies.

The most common method of intercepting run-off is to provide gullies at regular spacings at the edge of the carriageway, in the channel adjacent to the kerb.

Gullies may have several variations and either can be gratings and frames which are set just below the level of the adjacent road surface or can be set into the kerb and have an access cover at footway level. The latter type are generally considered to be less efficient at intercepting flows along the channel. If gully grating is to be set at a level lower than the adjacent carriageway, this should be no more than 6mm. Differences in level greater than this can represent hazards for drivers, cyclists and pedestrians.

Gratings can have either straight or curved bars. Straight bars are appropriate for use where the road longitudinal gradients are relatively flat. Gratings with curved bars should be used where the gradient is steep as they tend to be more efficient at intercepting flows along the channel where the velocity is quite high.

Gully gratings sit over a gully pot normally constructed from precast concrete, clay or plastic. To collect the accumulation of deposits from the road surface washed off during a rainfall event, it is normal to specify a trapped gully.

The gully pot acts as a catchpit and prevents the silt and other detritus from entering the highway drain or receiving watercourse. In urban areas, where there are combined drainage systems still in use, trapped gullies are essential to prevent smells rising from the sewer.

Gullies should be spaced to effectively intercept all run-off from the carriageway.

As a general rule of thumb, gullies should be spaced 40 to 50m apart or one gully should be provided per 200m^2 of drained area, although in 1958 Mollinson presented a paper to the Institution of Highway Engineers recommending that gully spacings should be calculated as follows:

$$D = \frac{280\ S^{0.5}}{W}$$

where

D = gully spacing in metres
S = gradient of channel expressed as a percentage (e.g. 1:40 = 2.5%)
W = width of paved area in metres

Nominal carriageway longitudinal gradient (1 in...)	Area drained to each gully
15	330
20	330
30	275
40	240
60	200
80	180
100	167
150	160
200	160

***Table 12** Highway gullies – suggested maximum drained areas*

Table 12 gives a guide to the area of carriageway which can be drained to a single gully and may be used where the carriageway has a consistent longitudinal gradient.

Where longitudinal gradients are very flat, ponding often occurs at junctions of side roads, particularly when the longitudinal gradient of the side road is also flat. In these locations, false crowns will be necessary to prevent water from lying on the carriageway surface and gullies should be sited at each tangent point immediately upstream of the junction.

Where there are false crowns between gullies, then the gully spacings should be no more than:

$$S = \frac{d}{2}\ \text{m}$$

where

S = gully spacing in metres
d = height of false crown in millimetres

Where a carriageway has superelevation, to prevent surface water from flowing across the carriageway, a gully should be provided at the upstream end of the cambered section. Additional gullies should also be provided at locations where the effects of surface water ponding on the carriageway may be inconvenient or may present a hazard such as at bus stops, upstream of pedestrian crossing points, lay-bys and on the upstream side of speed humps, speed tables and similar traffic calming features.

Double gullies are often specified at low points and this can be an economical way of draining the carriageway. On flat carriageways, gully spacings should be decreased to reduce the risk of ponding. On steep carriageways, due to the higher velocity of the run-off in the channel, there is a risk that even with curved bars, some of the run-off will flow over the grating to the next gully and so the gully spacings here should also be decreased, or double gullies provided at the normal spacing for single gullies.

Considerable research has been undertaken on this subject but the results are outside the scope of this book. The exact procedure for determining gully spacings and further information can be found in *Contractor Report 2 - the drainage capacity of BS road gullies and a procedure for estimating their spacing*, published by the Transport Research Laboratory.

7.3 Porous pavements

Porous pavements constructed with pervious macadam wearing course can be used to effectively reduce the amount of spray which occurs behind vehicles travelling at speed. This type of pavement is discussed in more detail in Chapter 11.

7.4 Run-off to channels

Depending upon the length of carriageway, the construction of slip-formed concrete channels at the roadside or in a central reserve is probably the most cost effective method of draining a carriageway. The critical points to consider are the depth of the channel (which should be no more than 150mm unless located within a protected central reserve) and the spacings of the outlets.

The channel can be constructed so that the outlets which generally take the form of open gratings or weirs are positioned directly above a drain and fall into a chamber through which the drain passes. The spacings of the outlets are determined by the drained area and the longitudinal gradient of the channel and should be positioned to ensure maximum efficiency.

When the outlets are positioned directly over chambers, the gratings can take the place of conventional manhole covers and can be designed to serve as either hinged or removable openings for man-entry into the chambers. Openings should be sized so that safe entry and exit is achievable.

7.5 Over the edge drainage

This method of draining the carriageway depends on adequate crossfall and allows surface water run off to literally drain over the edge of the carriageway where it can be intercepted by gravel topped French drains.

8

Foul drainage

8.1 Introduction

Foul sewage is made up of two components:

- Domestic sewage
- Trade effluent

Water consumption and foul flows are closely related, with the water being used in similar quantities to the foul flows being generated by its consumption.

Although metered consumption includes any water lost from the water supply system, the rate of infiltration of groundwater into the foul sewers can be estimated on average as being 10% of the flow in the sewer (new sewers) and 40% of the flow in the sewer (in the case of existing sewers) and this is approximately equal to the water lost from the water supply. As a result of this, when basing foul flows upon metered supplies, it is not generally necessary to make an allowance for infiltration unless a deduction is made for the waste from the metered water supply.

For accurate assessment of infiltration rates, late spring night time flow measurement should be used as the rate of infiltration depends upon the structural condition of the pipe and the height of the ground water table. Further, infiltration can be much higher than the actual foul flow.

Flows in domestic foul applications can be designed based upon a predicted flow of 4000 litres per unit dwelling per day, which is roughly equivalent to 200 litres per head per day (based upon three persons per dwelling) or can be sized using anticipated discharges into the sewer based upon the number of appliances connected to the pipes which lead into it.

Foul sewers are generally designed around dry weather flow (DWF). This is defined as being the rate of flow of sewage in a sewer during dry weather and includes infiltration. Flows in the sewer are normally based upon a multiple of the dry weather flow – usually 6 – and the total design dry weather flow is the estimated average made up of industrial flow, domestic flow, and infiltration plus an allowance for high usage premises such as hospitals and schools. The design DWF gives a prediction of flows for 20–25 years ahead.

It should be noted that the DWF multiplier should not be applied to infiltration because the rate of infiltration is generally constant over a 24 hour cycle. The multiplier is only used to take account of the diurnal variation in foul flows.

8.2 Domestic sewage

Domestic sewage is the term used to describe the discharge from domestic appliances such as toilets, washbasins, sinks, baths and showers. Although termed domestic, these appliances can be situated in places other than private dwelling houses such as offices, restaurants, airports, industrial units, hotels, hospitals and schools.

Flows from industrial areas are also included in domestic flow calculations. There are considerable differences in the rates of flow from industrial units, but metered consumption is often a reasonable guide as to the likely flows contributing to the sewer.

For flows from schools, an allowance of 50 litres per head per day; and for flows from commercial areas, an allowance of 110 litres per head per day per working person give fair indications of the likely flow contributing to the sewer.

8.3 Trade effluent

In addition to the effluents arising from industrial processes, examples of trade effluents include:

- compressor and boiler blowdown
- steam condensate and cooling water
- vehicle and plant cleaning effluents
- swimming pool backwash water
- air conditioning water
- pressure testing liquids

Trade effluent is the term used to describe any discharge which is produced by an industrial process. Its discharge into a public sewer must first be approved by the drainage authority which will issue a trade effluent licence. Discharge into a watercourse must be approved by the Environment Agency.

If foul sewage is left in any one place for any length of time, or is very slow moving, septicity and settlement of the suspended solids starts to occur. This leads to offensive smells and creates difficulties at the treatment works.

Septicity occurs when the bacteria which are present within the sewage can no longer respire aerobically (i.e. in the presence of oxygen) and anaerobic respiration takes place instead. This form of respiration is not a problem until the bacteria start to respire aerobically again, for instance where the flow in the sewer becomes faster or more turbulent or at the point of discharge of a rising main.

This happens because oxygen is introduced into the sewage and gases such as hydrogen sulphide (H_2S) and sulphur dioxide (SO_2) are released by the bacteria and these cause the offensive smells. In dilution, for example in the open air, these are not much more than bad odours, but in concentration these and other gases such as carbon dioxide and methane can cause asphyxiation and death if appropriate measures are not taken by operatives working in a live sewer. Methane is inflammable and explosive.

Further, when hydrogen sulphide gas (sulphurated hydrogen) is dissolved in water, sulphuric acid (H_2SO_4) is formed and this can attack the soffit of the pipe system. This is frequently evidence of septicity.

In well designed systems, the depths of flow will be such that settlement of solids does not itself present a problem, but in the design of trunk sewers, septicity may occur due to the time of travel within the pipe.

Long lengths of rising main are likely to result in septicity of the sewage and the location of any pumping station should take this into consideration so that the lengths of rising main are as short as practical and the lengths of gravity sewer are as long as can be economically constructed.

Chemical dosing and oxygen injection plant and techniques have proved to be worthwhile methods of overcoming the problems of septicity, but can be costly to implement.

Acidity and alkalinity often arise as a result of trade effluent being discharged into the sewer. Excessive amounts or strengths of alkaline or acidic sewage are likely to give rise to difficulties at the treatment plant and may be detrimental to the sewer material. For these reasons, aggressive trade effluents should undergo some form of treatment prior to their entering the sewer and this may be a requirement of the Sewerage Undertaker prior to issuing a trade effluent licence.

Acidity created as a by-product of septicity or directly from trade effluents from pickling, galvanising, wool and textile industries for example, or effluents which may contain inorganic sulphides, sulphates or other sulphur-containing substances can attack Ordinary Portland Cement (OPC), metals and cement mortar by reacting with the hydrated calcium aluminates or free limes in concrete or cement to form calcium sulpho-aluminate and this leads to the disintegration of the pipe from the inside outwards, ultimately affecting pipe density, impermeability and strength.

Other trade effluents may produce high concentrations of ammonium or metallic salts. Ammonium salts may affect the strength of the concrete over a period of time, especially where the concentrations are above 1%. Metallic salts may result in a precipitation which gives rise to corrosion by electro-chemical activity and affects pumps and metal pipework.

Trade effluents can also contain high proportions of grease, tar, oils, slurries and suspended matter which should be screened prior to discharge into the public sewer. Additionally, trade effluents may contain chemicals which on their own are relatively harmless but which may react with substances which are already in solution in the sewage flowing through the public sewer to form hard and insoluble encrustations on the inner surface of the pipes. They may also promote the growth of or contain living micro-organisms and bacteria which upon entry into the sewer become so prolific that blockages occur.

Domestic sewage can also contain high levels of grease, particularly when a sewer serves a restaurant community, hotel, or other building such as a hospital where large numbers of meals are cooked. Fast food outlets are prime contributors of grease in public sewers and many local authorities now insist on grease traps being installed for new developments such as these upstream of any new connections to the public sewer.

Accumulations of grease cause deposits to be formed on the inside of the sewer and this reduces its flow capacity resulting in other difficulties such as surcharge or flooding. Over time this reduction in capacity will lead to blockage and flows backing up, resulting in suspended solids remaining on the invert of the sewer and further reducing its capacity.

8.4 Foul sewer design

As with surface water sewers, foul sewers should be laid at gradients sufficient to achieve self cleansing velocity, otherwise solids will settle on the invert of the pipe, restricting flow and giving potential for blockage.

For calculating flows, dry weather flow is given by the following formula:

$$\mathrm{DWF} = PG + I + E$$

where

P = population (no)
G = consumption per head per day (cu.m/day or litres/head/day)
I = Infiltration (cu.m/day)
E = Industrial flow and trade effluent in 24 hours (cu.m/day)

For multiples of DWF it is only the PG that is factored, e.g.

$$6\mathrm{DWF} = 6PG + I + E$$

I remains as the infiltration at DWF and not 6DWF
E remains as the industrial flow at DWF and not 6DWF

For domestic flows, a figure of 4000 litres per unit dwelling per day can be used on residential estates and based upon this figure, ***Table 13*** gives the approximate number of units that can be served by various pipe diameters laid at different gradients.

Gradient	Pipe diameter (mm)			
	100	150	225	300
1:20	253	750	2210	4745
1:50	160	473	1396	2998
1:67	138	409	1208	2595
1:83	123	366	1080	2320
1:100	112*	334	985	2117
1:167	87*	258*	762	1638
1:200	79*	235*	695	1494

These figures are based upon a coefficient of friction, $k_s = 1.5mm$
** Pipes shown thus do not achieve the self-cleansing velocity of 0.75m/sec, even at pipe full flows*

Table 13 *Approximate numbers of dwellings per pipe*

As an alternative method of predicting flows, the discharge unit method may be used. This method should only be used at the a point just downstream of the stack which serves the appliances and takes the number of appliances (i.e. toilets, showers, baths, etc.) in a given building and assigns a probability of discharge to each of those appliances based upon the use of the building. Each of the appliances discharges at a known rate and is given a discharge unit value. The cumulative number of discharge units can then be converted into a rate of discharge, i.e. flow into the sewer.

9

Sewer rehabilitation

9.1 Introduction

Sewer rehabilitation techniques can be split into two categories. The first is the conventional method which has been used ever since the first pipe was laid underground and the technique is simple: excavate down to the defective pipe, repair it or replace it and reinstate the trench. There are times when this method is the only viable method of repair, replacement or new lay, but in congested streets it can cause significant disruption, not only to traffic but to pedestrians and shops/businesses and can have a direct impact on trade if open excavations are carried out in busy streets for long periods of time.

Open trenching methods should be used only for short sections of pipe replacement, repair or new lay and in locations where disruption is likely to be at a minimum, for example in back streets, estate roads and new development sites.

The second category or the alternative method is to employ trenchless techniques. These are methods which minimise the disruption associated with open trenching and the results are often speedier and more cost-effective.

Trenchless sewer and pipe repair systems can cope with a diverse range of situations, but each repair system is designed for its own specific range of conditions. It has applications in gravity as well as pumped and pressurised systems.

There are three principal categories to which trenchless techniques can be applied and each category requires specific information.

- Repair, rehabilitation and renovation. Specific information relating to the size, shape, route and condition of existing pipe or sewer is required along with any information on chambers and branches.

- New installations. Information is required on soil and groundwater conditions and the proximity and location of other adjacent mains, services, other apparatus and potential obstructions.
- On-line replacement. Information is required on existing pipe materials, size, shape, pipe bedding and/or surround and the alignment of the pipe relative to other services, apparatus and cables and the location of incoming or connecting pipes.

For rehabilitation and on-line replacement, the use of the CCTV camera is the primary requirement and tool used in these first investigations. Supplementary techniques to the CCTV camera include radar and sonar, with radar often being used in-pipe to detect the presence of external voids by penetrating the pipe walls. Sonar is often used underwater, for example in surcharged pipes and can be used in part full pipes in conjunction with a CCTV camera above top water level.

Stand-alone joint testing equipment is also available which tests the effectiveness and tightness of pipe joints but increasing use is being made of systems which perform the same tasks in conjunction with a CCTV camera.

For new installations, conventional site investigation techniques can be used to ascertain the existing site conditions and for on-line replacement, pipe and cable locating tools and ground penetrating radar systems are also used.

CCTV cameras can be inserted into most pipes from about 50mm diameter upwards and have a wide range of accessories such as skids and steering controls, to fit most pipe and conduit shapes and sizes, and with adequate lighting can be used in pipes up to 2000mm diameter.

Tractor units to propel cameras are generally confined to use in pipes over 150mm diameter and are remotely controlled from the surface. Most CCTV systems provide live feedback to the operator via a monitor.

CCTV cameras can be used to survey lateral connections from within the main pipe when a secondary demountable camera is carried on the body of the main camera and launched into the branch via remote control.

The use of ground penetrating radar (GPR) systems is becoming more commonplace and can be used to detect discontinuities below ground. GPR systems work best in dry, granular type soils and can reveal changes in strata such as road construction layers, voids and hard spots such as rock formations. The use of GPR in waterlogged soils and heavy clays is not always successful due to the limitations on signal penetration imposed by its frequency. A lower frequency gives better penetration; lower resolution and higher frequencies give better resolution but lower penetration. GPR is normally controlled and monitored from the surface and is similar in looks and operation to a hover mower. GPR can however, be used in pipe, in which case it is usually tractor- or skid-mounted, often in conjunction with a CCTV camera.

9.2 Trenchless techniques

Trenchless technology employs a wide range of methods to achieve its goal and the three principal areas mentioned at the beginning of this section have their own range of applicable techniques. ***Table 14*** gives a summary of the techniques and their applications.

9.2.1 Rehabilitation and repair

In this section, the following rehabilitation and repair methods are discussed:

- Sliplining
- Spray lining
- Close fit lining
- Cured in place lining
- Localised repair and seal

9.2.2 Sliplining

In its simplest form, sliplining can be described as the insertion of a new

Technique	Description/application
Sliplining	Installation of semi-rigid liner into defective (host) pipe
Spray lining	Liner sprayed onto inside face of defective pipe
Close fit lining	Installation of semi-rigid liner into defective (host) pipe
Cured in place lining	Installation of liner into defective (host) pipe
Localised repair and seal	Local repairs using local internal liners fitted into defective pipe
Pipe joint sealing	Local or whole pipe length joint sealing repairs
Robotic repairs	Remotely executed repairs controlled from surface
Sleeve/patch repairs	Local repairs using local internal liners fitted into defective pipe
Resin injection	Injection of resins into pipe surround to give added structural integrity and prevent leaking joints
Pipe re-rounding	Reshaping of pipes prior to lining
Pipe-bursting	Breaking out of defective pipes using impact mole prior to online replacement
Pipe-splitting	Splitting of defective non-brittle pipes prior to repair or on-line replacement
Pipe-reaming	Reshaping and resizing of pipes using a drilling head prior to relining
Pipe-eating	Reshaping and resizing of defective pipes using a microtunnelling machine prior to relining

Table 14 *Summary of sewer rehabilitation techniques and applications*

pipe into an old one. A pit is excavated at each end of the operation. One is a reception pit and may be used for pulling the new pipe through the existing pipe. The other is an insertion pit or 'starter trench' and may be used for jacking or pushing the new pipe into the existing pipe.

The most common material used in for the new pipe in sliplining is PE although in theory any material can be used. PE is flexible and can accommodate most of the minor bends encountered during normal installation and lengths are often fusion welded (butt-fused) together in a string prior to being installed in the existing pipe.

Sliplining is more often used in pressure pipelines as the resultant reduction in diameter can cause difficulties in gravity systems due to the subsequent reduction in pipe capacity.

Occasionally, the nominal bore of the existing pipe is slightly larger than the outside diameter of the new pipe and as a result, grouting of the annulus is required. Grouting of the annulus in pressure pipelines is a necessity when specifying thin walled non-structural liners as the new pipe will not have the necessary strength to deal with the operating pressures. These liners would be specified when the host pipe has sufficient restraint to withstand the internal pressures.

The grout may be high strength and add structural support to the liner, or it may act only as a filler and the existing pipe then provides the structural support. This latter technique is sometimes known as 'interactive lining'.

It should be noted at this point that sliplining in pressure pipes increases the friction head due to the reduction in pipe diameter and means reduced flows unless the pumps are uprated as part of the renovation scheme. This may lead to a requirement for new control panels and new supplies.

Thick walled pipes may be specified to withstand high external loads and internal operating pressures but consideration should also be given to grouting pressures and winching forces during installation. There may be a significant reduction in internal pipe diameter if a thick walled pipe is used and it may be more economical to specify a thin walled pipe with a high strength grout annulus to perform the same task.

Prior to insertion of the sliplining pipe, the individual pipe lengths must be jointed together and the joints must not come apart during insertion. Jointing can take place either on the surface or in the insertion pit.

With the former, long pipe strings can be laid out on the surface and the pipe can be winched or pushed into the existing pipe. The limitations of this are that for deep or large diameter pipes, the relatively large dimension of the tightest radius through which the pipe can bend before permanent deformation of the pipe occurs often means that the insertion pit will be quite long.

The second option allows for a shorter insertion trench but means that the insertion of the new pipe will only be as quick as the jointing operation of the sliplining pipes.

Jointing of the pipes is normally by fusion, but screw joints and snap fit joints are also used. Use of the former normally means that the sliplining pipe will be pulled into place from the reception pit, whereas pipes with either screwed or snap fit joints are often pushed into the host pipe from the insertion pit.

Another type of liner used in sliplining is the spirally wound liner.

There are two methods of placing a spirally wound liner into a host pipe. The first is by a hydraulically driven winding machine located in a manhole or other pit. The machine helically winds a uPVC strip into the host pipe and the strip travels as a tube, screw-like down the host pipe.

The uPVC strip is ribbed for strength with T-beams on the outer face and the whole tube rotates during installation. Frictional resistance within the host pipe is high and flotation may be used to reduce the load on the winding machine.

The second technique involves the use of a winding machine which travels through the host pipe, rotating within a winding cage shaped to suit the host pipe. Non-circular pipes can be repaired using this technique.

As with push/pull sliplining methods, grouting is carried out after insertion of the spiral. In large diameter pipes with the latter method of spiral winding, steel reinforcement can be inserted between the ribs of the liner for additional strength.

Expanded spiral liners have also been used successfully and have the facility to expand during installation to remove the requirement for a grouted annulus. This is a variation of close fit lining discussed later in this chapter.

Grouts are usually OPC and PFA (Ordinary Portland Cement and Pulverised Fuel Ash), although as with concrete, admixtures are available to increase or reduce viscosity (workability) and setting (curing) time.

When the grouted annulus is to provide added or sole structural strength or in applications where the liner is required to bond to the existing pipe, a compressive strength of 10–20kPa is generally required for the grout. When the grout is to act as a filler only, the grout will often only have a strength of 1kPa.

Flotation during the grouting operation is often a problem and it is common practice to avoid flotation by filling the pipe with water and grouting in stages. Flotation is inherently more of a problem in large diameter pipelines and the designer should remember that the flotation forces are related to the volume of the liner multiplied by the grout density and not just the mass of grout within the annulus.

Laterals and branch connections will usually be reconnected after the sliplining operation by excavation from the surface, although in large diameter pipelines, this can be carried out from inside the pipe.

9.2.3 Spray lining

Spray lining is a technique generally used to apply a protective coating to the internal face of a pipe. The application of this technique to non-man-entry sewers is in its infancy and is currently being developed. As yet there is no large scale commercial use of this technique. Currently, the most common use for spray lining is on potable water mains

The aim of spray lining and the purpose of sewer rehabilitation may be at opposing ends of the spectrum, as sewer renovation is usually required to increase a pipe's resistance to external loads rather than to prevent corrosion from inside.

Flows within a sewer need to be completely stopped during the application and curing of the sprayed lining and this may not always be practical, although there may be the benefit over other techniques that spray lining could remove the need for excavation in order to reconnect lateral branches.

Spray lining may rely on a bond with the host pipe and in this case, the internal face of the host pipe needs to be cleaned of rust, corrosion and scale.

Cleaning of the host pipe may involve high pressure jetting, cutters and flails, borers, scrapers and wire brush or foam pigs. Care needs to be taken to prevent damage occurring to the walls of the host pipe during some of the more aggressive cleaning methods.

There are two principal lining materials, epoxy lining and cement mortar.

Epoxy lining uses a resin which bonds to the prepared internal surface of the host pipe. This material has a shorter curing time than cement mortar linings and epoxy linings are generally applied as a much thinner coat. Defects in the epoxy coating may allow corrosion to continue and the resin itself offers no corrosion inhibiting properties.

The epoxy lining is applied by a spraying machine with a rotating nozzle drawn or propelled through the pipe. The speed of the machine and resin flow rate determines the thickness of the applied lining.

Epoxy resins are approved for use in the water industry, but should only be used for potable water mains when the resin formulation being proposed has been officially accredited by the water authority.

Cement mortar lining is a less expensive alternative to epoxy lining and is a commonly used material. The alkaline properties means that the mortar acts in a corrosion inhibiting capacity. The lining offers no structural strength but provides a smooth internal surface.

The application of cement mortar linings is generally via a spraying machine which has a rotating head and on small diameter pipes is usually hose-fed from the surface. Smoothing of the applied mortar is undertaken during a follow up trowelling operation, using rotating spatulas fixed to the spraying machine.

9.2.4 Close fit lining

Close fit lining is a technique quite similar in principle to slip lining, but the main difference is that in 'close fit' or 'modified sliplining' the pipe to be inserted is deformed prior to insertion and then reshaped once it is in position. Close fit lining has applications in both gravity and pressure pipelines and removes the requirement for annulus grouting.

There are two methods commonly used to deform the liner. The first is often termed 'swaging' and involves pulling or pushing the liner through a roller to temporarily and uniformly reduce the diameter of the liner. The rollers rely on the elastic properties of the liner material and there are therefore limitations to the amount of diametrical reduction that this process can achieve. It is best used on pipelines where there is little dimensional irregularity.

Swaged liners have been used on pipes from 100mm diameter up to 1100mm.

The second method of deforming the liner is sometimes known as 'fold and form' and involves folding the liner into a 'C' or 'U' shape prior to insertion and then using pressure and/or heat to restore its shape once it is in position. Folded liners have been used on pipes up to 1600mm diameter.

As with sliplining, one of the major drawbacks of close fit lining is that it reduces the internal diameter of the host pipe and may not always be suitable for gravity pipelines.

Once a close fit liner is in place, its shape is restored using either air pressure, in which case the two ends of the pipe are sealed and the internal air pressure is increased; or by heat, in which case hot water is passed down the pipe to soften the liner and thus allow it to return to its original shape. This latter method may be used in conjunction with increasing the internal pressure.

9.2.5 Cured in place lining

Cured in place lining is also referred to as 'in-situ lining' or 'cured in place pipe' (CIPP). As with close fit lining and sliplining, a liner is inserted into the host pipe.

The liner is generally a soft tube impregnated with epoxy or polyester resin which is expanded against the wall of the host pipe and is then cured, either at normal temperature (ambient cure), or by the application of heat, usually hot water or steam (thermal cure). Other systems are in use which are cured using ultraviolet light.

Once cured, the liner offers structural strength to the host pipe and unlike sliplining, liners used in gravity pipelines do not need to bond to the host pipe.

The inherent flexibility of cured in place liners means that they can be used for almost any shape of cross section of conduit, although for large non-circular sections where structural loadings are high, the necessary increase in liner wall thickness, weight and cost may make this a prohibitively expensive operation.

Some CIPP linings can be prefabricated to take into account changes in cross section between manholes and others can stretch to accommodate minor cross sectional variations.

Before CIPP liners can be installed, all debris, loose pipe fragments, intrusions and encrustations must be removed and the pipe cleaned. Flows in the host pipe must be prevented and surcharge pressures may cause the liner to set in a deformed shape if not properly controlled, although this is mostly evident in deep sewers.

Ambient cure systems tend to be less expensive than thermal cure options and are generally used in small pipelines up to 150mm diameter as the lack of control over curing makes them unsuitable for pipes over this size. The fabric of the liner is usually a polyester resin impregnated or coated felt, with impregnation often carried out on site.

During installation, the liner is pulled into the host pipe and a temporary inflatable inner sleeve is then pulled through it which is inflated using either water or air. The inflated liner remains in place until the resin impregnated liner has cured.

Thermal cure systems are generally more quality controlled as the impregnation of the liner is usually carried out under monitored factory conditions. The liner can be pulled into the host pipe and hot water used to fill and cure the liner, but quite often an inversion process is used instead, whereby water or air is used to turn the liner inside out and to push it along the host pipe.

Once the inversion process is complete, the water is recycled through a boiler unit and the water is heated and circulated through the inverted pipe. Temperatures are monitored using thermocouples and once the liner has cured, the water is cooled and released.

With both ambient and thermal cure systems, one of the most common causes of failure is premature curing of the liner. The curing process is exothermic and the temperature rise increases with the amount of resin in the mix. It is important that the temperature rise is controlled until the liner is in place.

Other liners are cured by their reaction to ultraviolet light and as with both ambient cure and thermal cure, a liner is placed in the host pipe. A tractor or skid mounted UV light source is then inserted into the liner, the ends of the liner are sealed and the liner is inflated. The UV light source is then moved through the pipe to effect a cure.

The liners used in all three systems are generally resistant to most chemicals normally found in sewers, although for particularly aggressive environments, specific inert or resistant resins can be formulated.

One of the main advantages of CIPP liners is that lateral connections can be cut and reinstated from within the main pipe and lateral relining systems are available which can be installed either from the host pipe or from the head of the lateral branch, although as yet there is no satisfactory method of jointing the lateral to the sewer when lining small diameter pipes. In small pipes, connections may be cut from inside the sewer and the lateral lined from the head of the lateral, but the joint cannot be sealed. There is a 'top hat' system currently available which achieves joint sealing, but its use is currently restricted to pipe diameters over 300mm.

9.2.6 Localised repair and seal

When damage is restricted to a short or isolated length of pipe, localised repairs and seals often provide the most cost-effective solution, particularly in non-man-entry pipes when the defects occupy less than 25% of the length of pipe, or where there are fewer than four defects each of less than 1 metre in length between manholes.

There are five main categories of localised repair system:

- Pipe joint sealing
- Robotic repairs
- Sleeve/patch repairs
- Resin injection
- Pipe re-rounding

9.2.6.1 Pipe joint sealing

As the name implies, pipe joint sealing is a system used to seal pipe joints. The method involves the installation of a metal band within the defective pipe.

For pressure pipelines, the external face of the band is usually coated with an elastomeric material and the band is expanded to form a seal with the surface of the host pipe. This system is available for joints in non-man-entry pressurised pipelines rated at up to 20 bar with diameters between 600 and 3000mm. Expansion of the band occurs at 2 to 3 bar pressure.

For gravity systems, the metal band is usually stainless steel which is expanded against the host pipe by the use of an inflatable packer. This is deflated upon completion and withdrawn and the metal band retains its shape due to an integral ratchet mechanism.

Pipe joint sealing in gravity systems using this technique may not always be effective, particularly as the ratchet may work loose allowing the band to move downstream.

9.2.6.2 Robotic repairs

Robotic repairs are used primarily in gravity pipelines of 200 to 800mm diameter. A tractor mounted or mechanically propelled skid mounted grinding robot is inserted into the defective pipe which removes encrustations and intrusions and prepares cracks for repair as it travels along.

The robot is fitted with a grinding head which if necessary, can cut through steel reinforcement, and it is this head which prepares the internal surface of the pipe for the subsequent repairs. Cracks are generally milled out to 25 to 30mm depth and width and protruding laterals are cut back. A second filler robot usually follows the grinding robot and this applies an epoxy resin to the prepared surfaces and around defects. The robots are controlled from the surface.

9.2.6.3 Sleeve/patch repairs

Sleeve/patch repairs often involve the use of techniques and materials similar to CIPP and close fit lining. An epoxy or polyester resin impregnated liner is pulled into the defective pipe and is then inflated with a packer or mandrel.

Ambient and thermal cure versions may be used and impregnation of the liner is almost always carried out on site and as with CIPP and close fit lining, premature curing of the liner is one of the most common causes of failure of this technique.

9.2.6.4 Resin injection

Resin injection techniques can be used for both local repairs and repairs to the whole pipe length.

For local repairs, an inflatable packer is generally positioned across a leaking or defective joint. The ends of the packer are pressurised to isolate the joint and the internal core of the packer is pressurised so that the pressure drop through the defective joint can be ascertained. A sealant can then be injected into the defective joint through the packer to seal the voids. The sealant is usually a polyurethane resin or an acrylic grout.

A second method of resin injection used primarily for stabilising the existing pipe structure uses an epoxy mortar or resin. In a similar manner to joint repairs, an inflatable packer is positioned to isolate the defective section of pipe and the epoxy mortar or resin is injected through the packer into the defect.

These two resin injection systems are very effective and can be used to fill large voids around pipes.

For repairs to a whole length of pipe, a 'fill and drain' system can be used whereby the defective section of pipe is isolated and a solution is poured into the isolated section and allowed to infiltrate into cracks, defective joints and voids around the pipe.

The solution is then pumped out and the pipe is filled with a second solution which is undergoes a chemical reaction with the first to form a gel. This technique is mostly used on large scale leakage reduction repair programmes due to economic considerations but has the advantage that it is a very quick and effective method of sealing defective pipes.

9.2.6.5 Pipe re-rounding

Pipe re-rounding is a technique which can be used to re-shape a deformed pipe prior to repairs being undertaken.

During pipe re-rounding, a metal clip is installed within the defective pipe and this holds the fragments in position until the pipe is repaired by the insertion of a CIPP liner or similar. The metal clip offers no structural support to the pipe and the repair operation should follow up the re-rounding as soon as possible.

To insert the metal clip within the defective pipe, a pipe expander unit is inserted into the defective pipe. The unit may be an inflatable mandrel type or may be a hydraulically operated steel mole with expanding rams.

During pipe re-rounding, the expander unit may take the line of least resistance and consequently may not always result in a successful operation, particularly if there is a void around the defect.

9.2.7 On-line replacement

On-line replacement involves the removal of a defective pipe and its replacement with a new pipe of the same or different diameter on the same line and level.

There are four main classifications to on-line replacement which are discussed in subsequent sections in this chapter:

- Pipe-bursting
- Pipe-splitting
- Pipe-reaming
- Pipe-eating

9.2.7.1 Pipe-bursting

Pipe-bursting is a technique which was developed in the 1980s and can be used on pipes up to 1200mm diameter, although most pipe-bursting operations are carried out on pipelines between 150 and 375mm diameter.

The pipe-bursting tool is generally torpedo-shaped and is pushed or pulled through the defective pipe. There are two types of pipe-bursting tool, one is hydraulically operated, the other is a percussive tool. They perform the same task but can have different applications.

Pipes used in pipe-bursting can be welded or snap-jointed polyethylene (PE) pipes or special clayware pipes designed specifically for pipe-bursting. The hydraulic pipe-bursting tool has an expanding head which in operation is pulled or pushed through the defective pipe to crack and split it. A new pipe string is then pushed into the old pipe on its original line, with the lead pipe of the new string being attached to the pipe-bursting tool. The tool is then pulled through the old pipe and the new string is simultaneously pushed into it using hydraulic jacks.

Hydraulic pipe-bursting systems should be used as an alternative to percussive pipe-bursting where the percussive action is likely to affect adjacent foundations or services for example and have successfully been used on pipelines up to 900mm diameter. Portable hydraulic pipe-bursting systems are available for use on small diameter pipes up to 150mm diameter where access is restricted, e.g. private gardens or under buildings.

Percussive pipe-bursting systems use a pneumatic mole housed within a steel cylinder. A pneumatically driven hammer strikes an anvil in the nose of the tool and this drives it forward, breaking the surrounding pipe as it travels along its length.

The percussive nature of the mole means that this method of pipe-bursting is best suited to brittle pipe materials such as cast and spun iron, concrete and clayware. There are many types of mole, some of which drive the broken fragments into the surrounding ground and simultaneously draw the new pipe string into the old.

9.2.7.2 Pipe-splitting

Pipe-splitting is generally used on pipes where the pipeline to be replaced is of a non-brittle material such as stainless steel, polyethylene or ductile iron. The system uses an expandable cutting head which cuts through the wall of the pipe and simultaneously draws a new pipe string in behind it and has been used successfully on pipes up to 300mm diameter.

9.2.7.3 Pipe-reaming

Pipe-reaming uses a directional drilling machine which grinds out the pipe walls as it travels along the defective pipe and simultaneously draws in the new pipe string behind it.

9.2.7.4 Pipe-eating

Pipe-eating can be used on existing pipe materials such as clay, concrete and GRP. A microtunnelling machine is drawn or pushed through the existing pipe and depending upon the cutting head fitted to the machine can increase the size of the bore and thus allow a larger diameter pipe to be installed as a replacement.

10

Pipe laying and materials

10.1 Introduction

In the UK, sewers are usually designed to set criteria laid down by the adopting authority or in the case of private sewers, to generally accepted standards.

There are two distinct methods of pipelaying: trenchless techniques and traditional open cut methods, i.e. trenches; and often one method will be more suited to a particular site than the other.

In the case of adopted (public) sewers, these are generally located within the public highway or within easements which pass through land in private ownership. In the case of private sewers, these tend to be laid to suit the existing or proposed site topography.

There are many different pipe materials available for use in the construction of sewers. For structures, concrete and brickwork are generally accepted, although for small manholes, inspection chambers and access fittings, plastic is generally preferred. In pipework, there is a range of materials that can be used, but for adopted sewers, that range is limited to that which is accepted by the adopting authority.

10.2 Pipelaying practice

Pipes should be laid uphill, starting from the downstream end and socket joints should be laid with the sockets facing upstream. This practice allows for any water to drain away from the working area. For adopted sewers, pipes should generally be laid with a minimum cover of 1.2m under roads and 0.9m elsewhere. These dimensions have arisen from requirements for the distribution of loadings on the pipelines, protection from frost, prevention of interference from other services and protection from farming practices. These dimensions may be relaxed for non-adopted pipelines.

During recent years, site practice has incorporated the use of lasers to assist in accurately laying pipes using traditional open-cut methods. The laser is usually located in a manhole at the downstream end of the pipe to be laid and is then aligned vertically and horizontally with the trench on the required gradient.

Any trench opened for the purpose of pipelaying needs to be at least 300mm greater than the outside diameter of the pipe. This is to allow the person laying the pipe to place a foot either side of the pipe and to allow adequate bedding and or surround to the pipe.

Items that should be considered during pipeline design can include the following:

- the incorporation of root barriers when constructing pipelines adjacent to trees or the specification of trees with small roots where pipes pass through landscaped areas
- the incorporation of a lean mix-concrete backfill where the pipe is laid with its invert below the level of the base of an adjacent foundation and when the centreline of the pipe is closer than 1m from that foundation measured horizontally
- installation of adequate trench supports during construction
- where the soil is unstable, consideration should be given to the type of material used for construction of the pipeline and for manholes
- where the groundwater table is high or tidal, preventative measures need to be taken to ensure that flotation of the pipeline does not occur

Pipelines constructed in close proximity to buildings, trees or in unstable ground will require additional precautions over and above those normally considered to mitigate the risk of damage to the pipe in use and to reduce the risk to the pipe laying operatives during construction.

Prior to pipelaying, trenches should be opened only as far in advance as is necessary and should not be left open for any significant duration. Once laid, pipelines should be tested prior to backfilling which again should be carried out as early as possible after the pipe has been successfully tested.

Depending upon the level of the water table, dewatering may be required when the trenches are opened. In most cases, diversion of the water away from the working area by the excavation of a trench and a sump from where the water can be collected and pumped away is normally sufficient. Where the water table is high and the excavations are deep, well-point dewatering systems may be required.

Where a pipe is to be laid at depths of greater than 6 metres, or in awkward locations, open trenching is not always the most economical or practical method. In cases such as these, the use of trenchless techniques come into play, with excavation in headings, thrustboring and tunnelling frequently used options. Headings can be constructed in rock and firm cohesive soils and should be backfilled as pipelaying proceeds.

10.3 Pipe bedding and sidefill materials

Any pipe laid in the ground needs to have adequate support and the type of support required is determined by the loads imposed upon the pipeline from soil and water above; traffic and wheel loadings; and construction traffic during construction.

There are two main categories of buried pipelines – flexible and rigid. A rigid pipe carries all the loads upon it without deformation. The bedding and sidefill materials enhance the loadbearing qualities of the material. Clay and concrete pipes fall into this category.

Flexible pipelines such as polyethylene and plastic pipes are designed to carry much less load and are allowed to deform to a certain degree. The bedding and sidefill materials in these cases provide a far greater degree of added structural strength. Bedding and sidefill materials should have the following properties:

- chemically durable and inert so as not to react with either the pipe material or the soil or groundwater
- easy to distribute beneath the pipe to allow adequate compaction by tamping
- easily workable

- a suitable, durable material that does not break up
- a rounded or angular material without sharp edges
- material of a suitable size
- the grading of the material should be such that it does not allow the migration of fines through it by the passage of water

The bedding required for the pipeline is also determined by the width of the trench. There are three types of bedding for pipes, and each type of bedding has a bedding factor, F_m. The higher the bedding factor, the greater the degree of protection offered by the bedding. ***Table 15*** gives a selection of bedding factors for different types of bedding.

Bedding factor F_m	Bedding Class	Description
4.8	A_{rc}	Reinforced concrete cradle or arch with reinforcement equal to 1% of area of concrete
3.4	A_{rc}	Reinforced concrete cradle or arch with reinforcement equal to 0.4% of area of concrete
2.6	A	Unreinforced concrete cradle or arch
2.2	S	Granular surround taken up to a minimum of 100mm above pipe soffit
1.9	B	Granular bedding taken up to mid diameter of pipe
1.5	F	Pipe laid on a flat layer of single sized granular material
1.5	-	Thrustbored pipelines
1.1	D	Pipe laid directly on hand trimmed trench bottom

Table 15 *Bedding factors*

Nominal pipe bore (mm)	Maximum particle size (mm)	Imported granular material	Class of bedding
Rigid pipes			
100	10	10mm single size	S, B, F
100–150	15	10 or 14 mm single size or 14 to 5mm graded	
150–500	20	10, 14, or 20mm single size or 14 to 5mm graded or 20 to 5mm graded	
>500	40	10, 14, 20 or 40mm single size or 14 to 5mm graded or 20 to 5mm graded or 40 to 5mm graded	
Flexible pipes			
100	10	10mm single size	n/a
100–150	15	10 or 14 mm single size or 14 to 5mm graded	
150–300	20	10, 14, or 20mm single size or 14 to 5mm graded or 20 to 5mm graded	
300–600	20	10, 14, or 20mm single size or 14 to 5mm graded or 20 to 5mm graded	
>600	40	10, 14, 20 or 40mm single size or 14 to 5mm graded or 20 to 5mm graded or 40 to 5mm graded	

Table 16 *Granular bedding and sidefill materials for pipes*

The three classifications of bedding are as follows:

- granular bedding/surround
- concrete protection
- trench bottom

Granular bedding and surround should consist of aggregates conforming to BS882, air cooled blast furnace slag to BS1047 or pulverised fuel ash to BS3797. Large diameter and heavy pipes require that the granular material is angular or irregularly shaped so as to provide better stability, although these require a greater compactive effort than rounded materials. ***Table 16*** gives details of the preferred granular bedding and sidefill materials for rigid and flexible pipes.

For detailed and simplified design of pipe bedding, reference should be made to the *Simplified tables of loads on buried pipelines* published by The Transport Research Laboratory.

10.4 Trenchless technology

There are three main categories of laying pipes using trenchless technology. With all three techniques, it is essential that adequate site investigations are carried out along the whole of the intended route, as unforeseen ground conditions are one of the most common reasons for failure. If inadequate or no investigations are carried out, there is a significant risk of specifying the wrong type of machine and this may mean that the machine can become stuck in the ground leading to potentially expensive recovery operations.

It must be noted that the use of the three trenchless techniques for gravity systems should only be used where the gradient is steep. This is because the directional accuracy has not yet been developed sufficiently and, as a result, the steering heads employed in these techniques are still very much dependent upon the ground conditions.

The three techniques can be categorised as follows and their applications have been summarised in ***Table 17***:

- Pipejacking and microtunnelling

- Directional drilling and guided boring
- Impact moling and pipe ramming

10.4.1 Pipejacking and microtunnelling

Pipejacking and microtunnelling can be used on pipelines from upwards of 150mm diameter and the two systems, although similar, have distinct differences. Both, however, offer cost effective solutions to pipelaying when weighed against traditional open cut methods of installation.

Technique	Diameter range
Pipejacking	150mm dia. upwards
Microtunnelling	150mm dia. upwards
Directional drilling	Up to 500mm dia.
Guided boring	Up to 500mm dia.
Impact moling	50 – 200mm dia.
Pipe ramming	Up to 2m dia.

Table 17 *Diameter ranges for trenchless pipelaying methods*

Pipes are laid between launch pits and reception shafts and are generally straight, although curved alignments can be achieved by the use of adjustable hydraulic jacks or gyroscopic devices to steer the drive in place of more conventional laser guided equipment.

Launch pits and reception shafts are sunk and constructed using conventional civil engineering methods and may be caisson type, sheet piled or segmentally lined. Where ground conditions are very good, unsupported shafts may be used. The shape of the pit may vary to suit the application and can be circular or rectangular. Launch pits must provide a suitable thrust wall for a jacking frame without affecting the structural integrity of the shaft.

With both pipejacking and microtunnelling, special jacking pipes are used which are specifically designed to withstand the forces brought about by the operation. In most cases, the pipe material will be either clay or concrete, but other materials such as steel, ductile iron and plastic are available.

Joints in jacking pipes are contained wholly within the pipe wall, unlike the spigot and socket type joints of conventional pipes and this allows the external surface of the jacking pipe to remain flush at joint locations and to offer less frictional resistance during the operation.

There are two main forces which need to be overcome during pipejacking and microtunnelling and these are friction and the mass of the pipe string being driven.

To overcome friction, the smallest diameter necessary to achieve satisfactory hydraulic design should be specified and the bore should be lubricated using a bentonite mud or bentonite/polymer mixture applied during the drive. The mixture is injected through holes drilled in the pipe walls into the surrounding ground and is often controlled from the surface by computer.

Pipejacking, as its name implies, involves the driving, or jacking, of a pipe so that a series of pipes is installed to form a continuous string in the ground. The jacking pipes are driven by hydraulic jacks located in a launch pit which houses a jacking frame and provides a stable thrust wall against which the frame can react.

The first pipe in the drive, known as the lead pipe, is fitted with a jacking shield mounted on hydraulic steering jacks and is lowered into the launch pit. The jacking frame exerts a horizontal force against the back of the lead pipe and the pipe is pushed forwards into the ground.

As the lead pipe makes progress, spoil is removed from within the pipe and further jacking pipes are lowered into the launch pit behind the lead pipe and are joined by integral flush fitting collars.

To steer the lead pipe on a correct course, adjustable hydraulic jacks are located within the jacking shield. They are generally controlled manually in pipe.

Removal of the spoil is often carried out by hand with a miner located within the lead pipe to excavate the ground ahead of the shield. Spoil is removed to the surface via a wheeled or rail-mounted mucking out skip behind the miner, or a conveyor belt system.

In difficult ground conditions, the miner can be replaced with machinery such as a cutting boom.

The two techniques described above are known as open shield methods and can only be used where the ground is self supporting. Where the ground is not self supporting, close face shields should be used and these employ a rotating cutter head in the jacking shield.

When using closed face shields, a spoil removal slurry is forced to the face of the excavation under pressure and this displaces the excavated material and drives it back along the excavated pipe to a suitable collection point.

As an alternative to the use of slurry as a method of spoil removal, the amount of excavation removed from the face may be restricted so that a sufficient level of excavated material remains at the cutting face to provide support. This technique is known as earth pressure balance.

Where lubrication alone is insufficient to allow a successful pipejack, interjacking may provide a solution. Interjacking may be used, for example, where the friction along the length of the pipe string is greater than the capacity of the jacking frame.

An interjack is a ring of hydraulic jacks set into its own jacking frame and located within a pipe string. The interjack uses the pipe behind to act as a thrust wall and it performs the same function as the jacking frame in the launch pit but breaks the pipe string down into more manageable lengths, each of which can be driven independently once the interjack is in position.

Microtunnelling involves the use of a long cylindrical machine with a rotating cutter head which is thrust in a bore ahead of the pipes. The outside diameter of the machine closely matches that of the external diameter of the jacking pipes. Microtunnelling is a similar method to pipejacking, and the operation could be described as mechanical pipejack.

Some microtunnelling systems are designed to be installed in small diameter launch pits of a nominal 2m diameter and can operate in difficult ground conditions, with the capacity to drive bores in soils where boulders are present up to one third greater than the diameter of the machine.

In use, the microtunnelling machine is lowered into the launch pit and is jacked against a frame to provide the forward thrust necessary to make progress and drive the bore. Spoil is removed through the machine by one of two methods, which are selected to suit the ground conditions.

In difficult ground conditions, loose soils and in locations where there is a high groundwater table, spoil is often removed using the slurry system as described earlier. The slurry is pumped through the machine to the excavation face where it mixes with the arisings and is recirculated back to the surface via a crusher built in to the machine. This ensures that no overly-large particles can enter and damage the slurry pumps.

In ground conditions where the soil is self supporting, an auger may be used to retrieve the spoil. An auger casing is located within the jacking pipe and feeds excavated material to a mucking out skip.

10.4.2 Directional drilling and guided boring

The terms directional drilling and guided boring are interchangeable, although differentiation between the two terms stems from the early days of development when directional drilling tended to be used for large and long distance drives such as highway and river crossings.

Directional drilling is used extensively for the trenchless installation of new ducts and pipes and may be used on alignments which are straight or gradually curved and can be steered around obstacles during the drilling operation. The cost of directional drilling is often significantly lower than traditional open cut methods of pipe laying even when ignoring the effects of disruption at the surface and environmental impact.

Unlike with microtunnelling and pipejacking, launch pits and reception pits are not always necessary and the boring machine can be set to drill into the ground directly from the surface, although surface-launch and pit-launch drilling machines are available. A typical mid-range surface launch machine is capable of installing a pipe of up to 500mm diameter over a distance of 100 to 350m.

In the past, directional drilling tended to be used solely for pressure pipelines, but advances in development mean that the accuracy of the drilling equipment has improved significantly and this technique can now be applied to gravity pipelines where ground conditions are suitable.

Installation of the pipe is often undertaken in two stages. The first stage is to drill a pilot hole along the proposed route. This hole is of a smaller diameter than the external diameter of the proposed pipe. The bore is then back-reamed towards the starting point to a diameter suitable for the new pipe, which is towed into place simultaneously with the back reaming operation. Where difficult ground conditions exist, percussive action drill heads can be used.

In operation, a drill head is pushed through the ground by a drill string. The drill head and the drill string both rotate and the head is lubricated by a drilling fluid such as a bentonite/water mixture.

A signal is transmitted from the drill head to the surface so that progress and drill head location can be monitored, giving the operator necessary information to steer the bore in the right direction. Wire guidance systems are also used, often when the bore to be drilled at a depth greater than the penetration depth of the radio-frequency transmitter or where the bore crosses a major river or highway.

The drill head exits at surface level or in a reception pit at a predetermined point, a back reamer is fitted to the drill string and the pipe string is also connected. The drilling operation then continues in reverse back towards the drilling machine, enlarging the pilot hole and drawing in the new pipe.

10.4.3 Impact moling and pipe ramming

Impact moling is one of the simplest of the trenchless techniques for laying small diameter pipes in straight lines over relatively short distances and is a common method of installing pipes under railway and road embankments.

The technique of impact moling is most suitable for short drives on pipe ranges from 50 to 200mm diameter. To minimise the risk of surface heave there should generally be 1m of cover for every 100mm diameter of tool.

Generally, impact moling can be carried out only in soils that can be compressed or displaced and because most moles have no guidance system, obstructions can mean that the bore will need to be abandoned. Impact moles can be deflected or stopped by obstructions and it is therefore essential that adequate ground investigations are carried out prior to any moling operations.

Monitoring of the mole's progress within the bore is possible by radio frequency detection, when a sonde is mounted in either the head or or rear end of the mole and the signal detected by a hand held scanner on the surface above the mole.

The moles used in impact moling are torpedo-shaped and percussive and are generally powered by compressed air, although hydraulically driven moles are available.

The percussive action of the mole drives it forward through the bore and this action can also be used to simultaneously draw the new pipe into the bore behind the mole.

Pipe ramming uses a percussive hammer to drive a steel casing to form a bore. The steel casing may be open or closed, depending upon the ground conditions. Open casings are preferred as they offer lower reaction against the ramming force and are less likely to be deflected by an obstruction. Pipe ramming is a non-steerable technique and drives generally are in the order of 50m with bores up to 2m diameter having been installed successfully.

In operation, pipe ramming is launched usually on a concrete mat set into a shallow launch pit. The lead casing is set onto the mat on guide rails and the ramming unit is attached to the rear of the pipe. The ramming unit drives the steel casing into the ground on a route determined by the guide rails until the end of the lead casing is close to the face of the launch pit. A second length of casing is then welded to the lead casing and the operation continues until the lead casing emerges in a reception pit. Casings are sometimes used as the main pipes e.g. for culverts but are more frequently used as conduits for mains or other services ducts.

10.5 Pipe materials

Pipe materials can be separated into two categories: rigid and flexible. Rigid pipes are pipes which are made from materials that offer structural strength to the pipeline and do not allow the pipe to deform under load. Flexible pipes do not have any significant structural strength and under load can deform significantly.

A range of the most prevalent rigid and flexible pipe materials and their applications are shown in ***Table 18***. Other pipe materials may include GRP (glassfibre reinforced plastic), RPM (reinforced plastic matrix), GRC (glassfibre reinforced cement), pitch fibre and asbestos. Most of these other materials, however are generally only found in existing drainage systems, although GRP is often specified in sewer rehabilitation works as a liner.

Rigid materials	
Clay	Readily available in diameters up to 600mm. Spigot and socket joints, 'O' ring joints and sleeve jointed systems are used. Available in standard strength, super strength and extra strength. Used usually in pipelines up to 300mm diameter. Not suitable for pressure pipelines.
Concrete	Readily available in diameters up to >2m. Spigot and socket jointed systems are used although on pipes for jacking socket arrangement is slightly different. Available in three strength classes, L (low), M (medium) and H (High). Used usually for pipelines over 300mm diameter where they become more economical than clay pipes. Not suitable for pressure pipelines.
Ductile iron	Available in diameters up to 1200mm. Spigot and socket joints/Tyton joints and flanged joints are most commonly used. Suitable for gravity and pressure pipelines
Flexible materials	
Polyethylene (PE)	Readily available in diameters up to around 600mm. Diameters over this size tend to be 'specials' and therefore expensive for short runs. Joints are usually fusion welded. Strength classes are PN10 and PN16 which indicate the nominal internal pressure rating for the material. Suitable for gravity and pressure pipelines.
UPVC	Readily available in diameters up to 450mm. Joints are generally spigot and socket type. Generally suited only to gravity pipelines and land drainage.

***Table 18** Rigid and flexible drainage materials and their application*

11

Roads

11.1 Introduction

As far back as the 5th century BC, the writings of the Greek historian Herodotus mention the highways built in Egypt for the purposes of transporting materials to and from the pyramids.

The earliest and probably the best known road builders in the UK of which there is still evidence were the Romans. Their early roads were generally made up of three layers of successively graded stones topped with a surface of stone blocks and were often up to 1.2m thick. Most of these roads disappeared during the Middle Ages.

Modern roads owe a great deal to two engineers, John Loudon McAdam and Thomas Telford whose pioneering roadbuilding skills came into existence during the first thirty or so years of the 19th century. The two methods of roadbuilding put forward by the two engineers were quite different.

McAdam favoured placing a layer of broken stone on a raised foundation of earth, with the belief that if the foundation was sufficiently drained, then the road should support any traffic. This system was adopted throughout Europe and was known as macadamisation, but after World War I this type of construction became less favoured as it was found that the roads constructed in this way could not carry the heavy trucks that were used during the war.

Thomas Telford's roadbuilding method involved laying a foundation of heavy rock which was raised along its centreline to facilitate drainage and then a surface layer of compacted broken rock was applied. This system was subsequently adopted for the construction of most roads.

There are several factors to be considered in modern road engineering. These include site topography which affects road gradients and alignments and determines the need for underpasses and bridges, cuttings and embankments; a prediction of the traffic loadings over the design life of the pavement; the ability of the pavement to support the predicted traffic loadings; and the pavement construction. Current standards for the UK suggest that the design life of a pavement should generally be considered as 40 years for a trunk road or motorway.

There are many different materials which go into the road make-up and these combine to make up four different classifications of pavement construction:

- flexible construction
- flexible composite construction
- rigid construction
- rigid composite construction

Flexible construction allows the loads imposed upon the pavement to be absorbed elastically. Materials used in flexible road construction include bitumens and asphalts. Flexible pavements comprise bituminous surfacing and bituminous lower pavement layers, whereas flexible composite pavements comprise bituminous surfacing and upper roadbase on a cement bound roadbase.

Rigid construction, as its name implies, does not behave with the same degree of elasticity as a flexible pavement. This type of construction uses concrete laid in slabs which can be reinforced or unreinforced. Rigid composite construction is an amalgamation of the two types of construction and comprises a rigid pavement layer overlaid by a flexible layer.

Each of these types of construction is discussed more fully later in this chapter.

Historically, traffic has been defined in terms of a cumulative number of standard axles to be carried over the life of the pavement. This is determined from an estimate of the number of predicted commercial vehicles per day calculated to be carried over the design life of the pavement.

The design of any new pavement needs to commence with a study of the traffic it is likely to carry. For residential and industrial estate roads, there are often county standards set down which will apply for consistency throughout the county, irrespective of the likely traffic; however, for new roads such as dual carriageways, relief roads, motorways and the like, the predicted traffic flows need to be assessed so that a reasonable design life for the new road can be determined.

Currently, a road pavement is constructed in layers. In its simplest form, the lowest layer will be the sub-base which distributes the traffic loads throughout the formation. Above the sub-base is the flexible, rigid or modular pavement construction. Where the formation does not have sufficient strength to support the pavement and predicted traffic loadings, a sub-grade improvement (or capping) layer is often laid beneath the sub-base. Other ground improvement techniques have also been used successfully and these include dynamic compaction and cement/lime stabilisation.

11.2 Pavement design

There are two approaches to the design of pavements: empirical and analytical. Current practice is generally to use empirical methods, that is, methods based upon experience of what has worked in the past.

There are four main principles to pavement design:

- ensure adequate drainage of the subgrade
- ensure no cracking of the roadbase will occur
- ensure that the roadbase is designed to spread the traffic loads adequately
- ensure that any permanent deformation, or that the cumulative effects of any permanent deformation in each construction layer, is not excessive

Without adequate drainage, the predicted life span of the pavement is significantly reduced. For this reason, the water table beneath the road should not lie close to the subgrade (600mm is the generally accepted minimum distance between the two). In order to achieve this, the installation of side drains parallel to the carriageway will lower the water table beneath the pavement. Further, as the pavement ages, the possibility of water ingress increases. The pavement should be designed so that any water falling on its surface is shed to one side of the pavement as quickly as possible and removed.

In addition to the above, during construction, water and moisture can enter the subgrade and sub-base materials. Effective drainage is therefore a necessity if this water is to escape without having any adverse effects on the pavement.

Roadbases should be designed to transfer the traffic loads adequately to the sub-base without cracking. Pavement failure occurs over the life of the pavement in service and if the roadbase is under-designed, then failure may not become evident at surface level for a number of years. Pavement serviceability changes with time and traffic. The critical life of the pavement is the point at which serious deformation begins and is defined in the UK as a rut having a depth of more than 10mm or as the onset of longitudinal cracks forming in the wheel track. Loads are applied to the pavement through the contact between the tyres of the vehicles using it and the wearing course.

This loading varies depending upon the speed and number of vehicles using the pavement. The roadbase must be designed to support these loadings.

Each commercial vehicle is related to a standard axle. This is defined as an axle which exerts a force of 80kN. This loading may then be said to equate to a particular number of equivalent standard axles and this number is termed the wear factor. The wear factor varies according to the type of vehicle.

This factor is required because the wear caused by an axle increases significantly with load and approximates to the fourth power law, meaning that a doubling of the axle load will be equivalent to a 16-fold increase in the wear caused by that axle.

11.3 Preparation and construction

The subgrade is the foundation upon which the road rests and can be the naturally occurring underlying strata or it can be fill material imported to make up the level or to replace an existing weak or unstable subgrade. Prior to any construction taking place, therefore, it is essential that the subgrade is assessed for strength so that appropriate thicknesses of materials can be specified.

The bearing capacity of the subgrade is usually measured in terms of the California Bearing Ratio (CBR). This is an empirical test derived by the California State Highways Department for the evaluation of the strength of subgrades. The CBR test relates all materials to a well graded, non-cohesive crushed rock which is considered to have a CBR value of 100%.

After many years in use beneath a road, the subgrade becomes well consolidated. The subgrade must be able to withstand sustained traffic loadings without undue and excessive deflection and this is controlled by the vertical compressive stresses at formation level.

The process of laying and compaction of materials is of prime importance and the underlying construction (the subgrade) needs to be of adequate strength and stability to receive the pavement layers and to allow for them to be compacted and trafficked.

Adequate compaction of the subgrade or of the layers above it will never be achieved if the subgrade moves about under a roller. In some cases a lowering of the groundwater table by the incorporation of sub-surface drainage will be adequate to allow satisfactory compaction, but in other cases more substantial subgrade improvement will be required.

The stiffness and strength of the subgrade are influenced by the moisture conditions which exist under the pavement. It should be ensured therefore that wherever a high water table is present during construction, the water table post-construction remains at a level no less than 300mm below the level of the formation under the pavement.

Subgrades which have been allowed to become excessively wet during construction reach lower equilibrium CBR values than those which have been adequately protected and drained. The ability of both the subgrade and the sub-base to carry construction traffic significantly decreases during severe wet weather and the adverse effects of trafficking a softened and wetted subgrade and/or sub-base will result in premature failure of the road.

Inadequate drainage can also have an adverse effect on the stability of the granular sub-base. Lateral groundwater flows can wash out the fines and this can result in the formation of voids in the sub-base material. This reduces the load transferring ability of the sub-base, can cause local collapse and deformation of the sub-base and will again result in premature failure of the road.

During construction, the subgrade should be shaped to the required profile prior to laying any materials above its surface as excessive variations and irregularities in the surface of the subgrade will lead to variations in compaction of subsequent layers and ultimately consolidation of those layers under trafficked conditions.

The in-situ CBR of the subgrade will vary with seasonal variations in moisture content and therefore it is usual to determine the design CBR from the plasticity index of from laboratory CBR tests on remoulded samples at moisture contents which simulate the long term equilibrium moisture content under the finished road.

11.4 Capping and sub-base layers

Under new roads, when the subgrade is found to be weak (normally where the CBR value has been found to be 5% or less), a subgrade improvement layer is usually placed on top of the subgrade to provide added strength as a foundation for the road. This additional layer is also known as a capping layer and can also have as its function the protection of the subgrade from the adverse effects of weather. The subgrade improvement material often consists of a local, low-cost material and there is a wide range of materials which have been used in capping layers, often comprising crushed rock or concrete.

There are certain requirements for the grading of this material, particularly if the road is to comply with the provisions of the *Design Manual for Roads and Bridges* (DMRB) in which case it is normal to specify a granular (type 6E, 6F1 or 6F2) material. Type 6E material is a granular material, used with cement for stabilisation of the subgrade. Types 6F1 and 6F2 are granular materials which are placed directly onto the subgrade and compacted. Type 6F2 is a coarser grading of type 6F1. Once subgrade materials have been laid, the target CBR values should be no less than 15% for the compacted layers.

For the sub-base, it is again usual to specify material from the *Design Manual for Roads and Bridges*, in this case an unbound granular (Type 1 or Type 2) material or a cement bound (CBM) material.

Type 1 granular materials include crushed rock, slag, concrete and well burnt non-plastic shale. Type 2 materials are similar and include those for Type 1 and also natural sands and gravels. Unbound granular materials such as these are not suitable for use under concrete carriageways. In construction, Type 2 granular materials perform better when laid in dry weather. They have a lower specification than Type 1 materials and do not generally give such a good performance when laid during wet weather.

There is an upper limit to the recommended depth of sub-base and this is generally accepted as 350mm, except in the case of weak subgrades where an additional layer of sub-base or capping should be laid first and compacted independently of the design thickness sub-base.

The sub-bases mentioned above are known as unbound sub-bases and are standard for flexible and flexible composite pavements. Cement bound materials (CBM) can also be used and are inherently suited for construction during wet weather and are standard for use as sub-bases during the construction of rigid and rigid/composite pavements.

It should be noted that cracking will often occur in cement bound sub-bases and this is due to the stresses caused by construction traffic and the temperature of the CBM material.

Lean concrete should be used and the sub-base must not be over strength.The extent of this cracking will generally be determined by the amount of construction traffic and the thickness of the sub-base and also the thickness and strength of the subgrade and/or capping layers below.

CBM materials are categorised in the Design Manual for Roads and Bridges and are referred to as CBM1, CBM2, CBM3 and CBM4 materials. Acceptable materials for use in CBM sub-bases are only partially specified and so a wide range of materials may be used, but only for CBM1 and CBM2 sub-bases and only as long as they comply with the requirements of the DMRB specification.

For CBM3 and CBM4 sub-bases, however, permitted materials are restricted to natural aggregates complying with BS882, air cooled blast furnace slag or crushed concrete complying with BS882. As an alternative to roller-compacted CBM sub-bases, there is a range of four cement bound sub-base materials which are compacted using vibratory methods. These are known as wet lean concrete 1–4. Wet lean concrete 1 may be specified for use under flexible and flexible/composite pavements.

11.4.1 Subgrade improvement and sub-bases

Where the subgrade is weak, improvement of the subgrade is required so that a solid and stable foundation for the road is achieved prior to any new construction or road layers being formed.

Unbound granular materials such as Class 6F1 and 6F2 are materials commonly used in capping the subgrade, often in depths of up to 600mm and it is necessary to ensure that any material used in the subgrade improvement is not frost susceptible. The depth from the final road surface that non-frost susceptible material should extend is dependent upon the frost index. Additionally, where the materials have a high sulphate content, their use should be restricted to away from areas where concrete is present due to the likely attack on the concrete by the sulphates.

11.4.2 Cement/lime stabilisation

Stabilisation and improvement of the subgrade by the addition of lime or lime and cement is a common treatment, but is not suitable for all soil types. Lime stabilisation is a good ground treatment for large areas such as car parks, and roads over soft ground. The associated costs of mobilisation are often high and therefore stabilisation may not be the most economical treatment for small sites. Stabilisation by this method has five main benefits:

- results in an increase in the shear strength and bearing capacity
- fast reduction in water content improves workability, strength and compaction characteristics
- improves the resistance to weather and construction traffic
- reduces the soil's susceptibility to swelling and shrinkage
- reduces the need for imported granular materials and the disposal of excavated material

When lime is added to a clay soil, hydration takes place and heat is generated. This reaction begins immediately. In their normal, unstabilised condition, the clay particles are surrounded by a layer of water but when lime is added, a chemical reaction occurs which results in ion exchange between the clay and the lime and this reaction reduces the size of the layer of water, causing the clay particles to move closer together and thus altering the structural properties of the clay. The principal changes are as follows:

- a reduction in plasticity and an increase in the plastic limit
- increased cohesion of clay particles resulting in a greater shear strength and also bearing capacity
- an increase in the optimum moisture content which allows better compaction at higher moisture content levels

Lime and cement/lime stabilisation can be used in many applications.

Wet PFA can be improved simply by the addition of lime. If cement is also added, then the strength gain of the PFA can be speeded up significantly, particularly during cold weather.

Where sites are overlaid with PFA fill, deep mixing of the PFA with lime or cement/lime can improve the site by reducing the pore water pressure in the PFA and hence allowing consolidation and better stability of the fill material.

Wet or cohesive soils can be treated with lime to form a good capping layer which will often achieve a CBR value of 15% or more. Where site-won materials are to be treated in this way, there is an obvious reduction in the need for imported suitable material. Further, once stabilised, the cohesive soils become impermeable, protecting the formation from the effects of weather.

Cement and lime can also be added to cohesive soils and chalks to provide a sub-base. Soils treated in this way are designed to replace sub-base material such as Type 1 granular material and can be directly overlaid with the pavement construction.

Treated soils offer good flexural and tensile strength characteristics and are usually impermeable. This means that they can be suitable to receive construction traffic.

Waterlogged sites can also be treated by the addition of lime. The reaction occurs immediately, generating heat and causing evaporation. Further moisture is absorbed during the chemical reaction. Depending upon the nature of the soil being treated, the treated material may then be used in general earthworks.

11.4.3 Alternative materials

There are several alternative materials that can be used in capping layers. They can be provided as purposely manufactured materials; as by-products of industrial processes such as from the iron and steel industry and the power industry; or may be naturally occurring materials. This last source provides the greatest percentage of materials used in the industry. Whenever alternative materials are proposed, it is often due to a shortage of conventional fill or because the materials are available locally. Alternative subgrade improvement materials may include the following, but some may not be accepted for use under adopted roads by the adopting highway authority:

- burnt colliery spoil
- china clay waste
- power station waste
- blast furnace slag
- slate waste
- construction and demolition waste
- spent oil shale
- incinerator ash

11.4.3.1 Burnt colliery spoil

Most burnt spoils are susceptible to frost. Burnt spoils are permitted for use as a granular capping material provided that they meet the relevant requirements of the Specification for Highway Works. Burnt spoils can be stabilised with cement to form a stabilised capping material, but as it is acceptable to use these materials in their unbound forms, there is little reason to do so. It is unlikely that burnt colliery spoil will be found to be suitable for use as an unbound sub-base material principally because it is doubtful that it could achieve the required 10% fines value (TFV) of 50kN, although colliery shale is acceptable as a constituent in cement bound materials CBM1 and CBM2 in both unburnt and burnt forms.

11.4.3.2 China clay waste

China clay sand has the potential to be used as an unbound granular capping material and has been successfully used as a granular sub-base. This material can be readily stabilised with cement to form a cement stabilised capping material.

As a type 2 sub-base (Specification for Highway Works), natural sands and gravels are permitted and therefore china clay waste can be used as an unbound Type 2 material, although it may be marginally frost susceptible.

China clay sand can be stabilised with cement to form an acceptable CBM1 and CBM2 cement-bound sub-base material.

11.4.3.3 Power station waste

There are two main types of power station waste: pulverised fuel ash (PFA) and furnace bottom ash (FBA).

Of the two, PFA is eminently more suitable as a capping material when stabilised with cement. It can also be stabilised with lime, but this falls outside the scope of the Specification for Highway Works. On its own, PFA is not considered to be suitable as a capping material and nor is it suitable on its own as a sub-base material, although it has self-hardening pozzolanic properties.

When stabilised with cement, PFA can meet the requirements for CBM1 cement bound-sub-base material. PFA can also be used as an additive to cement bound materials and PFA-modified concrete may be a suitable alternative to lean concrete (CBM3) material used in roadbase construction. FBA is a granular material but the particles are of a porous nature and therefore may not have adequate strength characteristics for use in road construction other than for minor roadworks.

11.4.3.4 Blast furnace slag (air cooled slag)

In its unbound form, blast furnace slag may be used as a granular capping material, except where there is a high water table or it is used in proximity to metals. It is also suitable for use as a granular fill and granular sub-base material. This is because below water level, blast furnace slag may leach out alkaline chemicals and there is an increased risk of groundwater pollution if this material comes into contact with metals that corrode in an alkaline environment. Blast furnace slag has an inherent self-cementing action and this makes it ideally suited for use as a capping material, although if required it can readily be stabilised with cement. The drainage characteristics of blast furnace slag make it unsuitable for use as a commencing layer below PFA.

11.4.3.5 Slate waste

Slate waste is suitable for almost all applications where crushed rock has been specified. It can be used as a granular capping material and a granular sub-base material when graded to comply with the specification for sub-base material.

11.4.3.6 Construction and demolition waste

Of the four categories of materials under this heading, i.e. bituminous materials, crushed brick, crushed demolition debris and crushed concrete, it is probably only the first and last which are of relevance in road construction, and their use as a sub-base or capping material is subject to their compliance with the grading requirements for such materials.

11.4.3.7 Spent oil shale

Spent oil shale is not unlike burnt colliery spoil and has similar characteristics. It can be used in almost all circumstances as an alternative to burnt colliery spoil.

11.4.3.8 Incinerator ash

Incinerator ash has properties which make it generally unsuitable for use in either bound or unbound forms as a capping material or as a sub-base material, although some ashes may be suitable for use as an unbound granular sub-base, although they are more suited for use as selected granular fill or bulk fill, but may present a contamination risk due to the potential existence of heavy metals.

11.5 Flexible construction

A flexible pavement construction consists of several layers of different bituminous mixes which combine to form a strong and durable pavement. The term 'flexible' is used to describe the way they absorb the loads and stresses imposed by the traffic above and the term 'pavement' is used to describe the layers which make up the road from formation level to finished road level.

The design life of a flexibly constructed road can often be extended by the application of a strengthening overlay rather than a need for complete reconstruction after the expiry of the design life period. Generally, there are four layers which make up a flexible pavement. From the ground up these are:

- sub-base
- roadbase
- basecourse
- wearing course

11.5.1 Sub-base

The sub-base is a layer of material which is placed on top of the subgrade (or capping layer) and has three main functions:

- to distribute and spread the wheel loads from the traffic above

- to provide an adequate thickness of frost resistant material
- to provide a working platform on which the roadbase and subsequent surfacing can be laid

11.5.2 Roadbase

The roadbase is the principal component of strength of the road and its main purpose is to absorb and distribute the loads from above to the sub-base. The roadbase, therefore, must not crack under the influence of traffic loadings. This cracking is induced by the horizontal stresses and strains at the bottom of the roadbase layer and is controlled by the strength of the material.

11.5.3 Basecourse

The basecourse is the lower part of the road surfacing and is also known less commonly as the binder course. The basecourse is subject to the highest stresses – shear and tensile forces – from the action of the traffic above. It contributes to the strength of the pavement and also provides an even and well regulated surface upon which the wearing course can be laid.

11.5.4 Wearing course

The wearing course forms the running surface of the road and is designed to withstand the direct effects of the traffic. It provides an even and weather-resistant surface for the traffic with appropriate and durable levels of skid and deformation resistance.

11.6 Flexible composite construction

Flexible composite construction comprises a cement bound sub-base overlaid with a base course and a wearing course, and as with flexible construction the road base and the wearing course form the upper layers.

11.7 Rigid construction

When considering rigidly constructed pavements, a similar process is followed to that adopted for flexible pavement design. The predicted traffic flows, the design life, subgrade, sub-base and concrete slab are all considered in turn. As with flexible pavements, the subgrade must be compacted and shaped to the required specification and improved or capped as necessary prior to receiving a sub-base.

Concrete slabs may then be laid and can be reinforced or unreinforced and their thickness will be determined by the predicted traffic flows. Joints in the slabs will be required but are also influenced by the time of laying.

For slabs laid during the warmer months (mid-April to mid-October), expansion joints may be replaced by contraction joints. Where limestone aggregate is used, the joint spacing may be increased by 20% due to the aggregate's greater resistance to thermal expansion. There are four types of joints used in concrete slabs and these are identified in ***Table 19***. Any reinforcement used in slabs should have a minimum of 60mm cover wherever practicable although in slabs less than 150mm thick, this may be reduced to 50mm.

Joint type	Description
Contraction joint	Allows the slab to contract during cold periods
Expansion joint	Allows the slab to expand during warm periods
Warping joint	Acts as a hinge in the slab
Longitudinal joint	Generally parallel to the alignment of the carriageway and perpendicular to other joints

Table 19 *Joint types in concrete slabs*

Where subgrades are particularly weak or unstable, a continuously reinforced concrete base with a flexible surface is often the most effective and trouble free solution as it minimises the risk of uneven settlement caused by reinstatement. In this form of construction no transverse joints should be used except for unavoidable construction joints. The surfacing should consist of two courses minimum (basecourse and wearing course) and should have an overall thickness of no less than 90mm.

11.8 Rigid composite construction

Rigid composite construction comprises a rigid pavement with an overlay of bituminous surfacing which forms the running surface. Due to its relatively high cost, its use is generally restricted to large projects such as trunk roads and motorways.

11.9 Pavement materials and their properties

There are four principal types of materials used in flexible pavement construction:

- asphalt
- coated macadam
- pervious macadam
- asphaltic concrete

11.9.1 Asphalt

Asphalts are mixtures of mineral aggregates, fillers and bitumen which distribute the stresses from traffic loadings through the mixture. To adequately resist deformation, the mixture must therefore have quite a high stiffness and in practice, a high filler content and hard bitumen is used to achieve this. There are two types of asphalt currently used in the UK: mastic asphalt and hot rolled asphalt (HRA).

Mastic asphalt is used primarily in specialist applications such as the waterproofing of structures, in locations where resistance to high loads is required and as footway surfacing. It is generally laid by hand at high temperatures using a wooden float. It has a high proportion of fines and is often laid in thicknesses of up to 50mm.

The skid resistance of this type of asphalt is low and there are several methods of increasing this resistance. Where it is used in carriageway applications, it is normal to allow the mastic asphalt to cool to a plastic state and then apply a layer of pre-coated chippings to the surface. Embedment of the chippings is secured using hand or light power rollers. In footways, the surface may be crimp rolled or treated with a special sand.

Hot rolled asphalt is used widely throughout pavement construction in the upper pavement layers. One of the most important characteristics of hot rolled asphalt is that it is gap graded; that is, it contains a very low percentage of medium size aggregate.

Where hot rolled asphalt is used in roadbases and basecourses, the bitumen content of the mixture is generally lower as the stresses from the traffic loadings are distributed through both aggregate interlock and the mortar. There are four main constituents of HRA and these are:

- filler
- fine aggregate
- coarse aggregate
- bitumen

In HRA, the filler is added to produce a mix that is more dense, and together, the filler and bitumen bind to form a mortar. The properties and characteristics of the mortar are dependent upon the amount of filler and bitumen viscosity.

Fine aggregate is the major constituent of the mortar and influences the performance of the mortar both in use and in application and may be made up from natural sands and/or rock fines.

Coarse aggregate is used to increase the volume of the mortar, thus making it more economical. Coarse aggregate also increases the stability of the mixture.

Bitumen acts as a binder for the filler and aggregates. Low penetration grade bitumens are used in heavily trafficked areas. High penetration grade bitumens are use where trafficking is less onerous on the pavement.

BS 594 provides guidance on the selection, design mixes and laying of hot rolled asphalt.

11.9.2 Coated macadam

As in HRA, there are four main constituents of coated macadams:

- bitumen
- filler
- fine aggregate
- coarse aggregate

As before, the bitumen binds the mixture. The filler fills the small voids in the matrix of aggregate and increases the viscosity of the binder. The coarse aggregates provide the main source of aggregate interlock which distributes the traffic loads throughout the pavement layers; and the fine aggregate partially fills the voids in the interlocking coarse aggregate.

Coated macadams can be broken down into four types, each with different applications:

- heavy duty macadam
- open graded and medium graded macadams
- dense and close graded macadams (includes DBM)
- fine graded macadams

BS4987 provides guidance on the selection, design mixes and laying of coated macadam.

11.9.2.1 Heavy duty macadam (HDM)

Heavy duty macadam is based upon a traditional dense bitumen macadam roadbase material, but to cope with the additional loads and stresses imposed upon a pavement by current and predicted traffic flows, the traditional roadbase macadam has been developed to provide better deformation resistant properties. HDM incorporates a hard grade of bitumen (50 pen) and a high filler content (7%–11%) and these two factors increase the stiffness by a factor of up to 3.

This increase in stiffness allows a reduction in layer thickness or an increased life of pavement.

11.9.2.2 Open graded and medium graded macadams

These macadams are used primarily in basecourse and wearing course mixes. They have a low fines content but after compaction the voids content remains fairly high at 15%–25% and when used as a wearing course, in order to make the surface waterproof a surface dressing is required.

The main strength of this material is brought about through the aggregate interlock; the bitumens used are generally 200 or 300 pen or cutback bitumens, depending upon the traffic loadings and the time of year during laying. For light and medium duty traffic applications such as car parks, drives and playgrounds, open graded and medium graded macadams provide adequate surface texture in terms of skidding resistance.

11.9.2.3 Dense and close graded macadams

Dense bitumen macadam (DBM) and close graded macadam are used in roadbase, basecourse and wearing course mixes but differ from open graded and medium graded macadams in that the fines content is high. This gives the material good load spreading characteristics and a high resistance to deformation.

These mixes do not have long term durability and are only suitable for light and medium traffic loadings, i.e. those that will not induce high stresses in the pavement construction such as on the spot turning by heavy goods vehicles, heavy braking, and high volumes of traffic.

Skid resistance is also lower than that required for higher speed applications.

11.9.2.4 Fine graded macadams

These macadams are normally manufactured using 200 pen, 300 pen or cutback bitumen. Fine graded macadams have a high void content which reduces as the material is trafficked and is often used in footway construction.

11.9.3 Pervious macadams/porous asphalt

Pervious macadams were developed in the mid-1950s from trials of a 10mm open textured macadam used in airfield applications to overcome the difficulties of aquaplaning and skidding on runways.

The surfacing became known as airfield friction course and was developed into a 20mm pervious macadam for use in road pavement construction.

These macadams can significantly reduce spray and noise and although using standard unmodified binders, the structural properties of pervious macadam are far less than hot rolled asphalt or dense bitumen macadam (a 40mm thick layer of pervious macadam is approximately equivalent to 20mm or 16mm of DBM or HRA respectively). The incorporation of a polymer modified binder or an epoxy bitumen substantially increases the stiffness of the macadam to that of HRA.

11.9.4 Asphaltic concrete

Asphaltic concrete was developed in the United States of America for use in aircraft pavements. In the UK its use tends to be restricted to airfields predominantly for two main reasons: firstly its lack of surface texture which means that it requires surface dressing and secondly, it requires a stiff supporting structure below it to prevent it cracking in use.

12

Bitumen technology and testing

12.1 Bitumen technology

Bitumen is a natural derivative of crude oil and has been used as a waterproofing and bonding agent since about 4,000 BC. Currently, world bitumen production is in the order of 20 to 25 million tonnes per annum, most of which is used in road construction and approximately 30% in Europe.

In bitumen manufacture, the crude oil is transported to a refinery where it undergoes two distillation processes. After these two stages of refinement, bitumen is produced in a form suitable for modification.

Over the last five years, the performance requirements for bitumens have become increasingly onerous and this has led to extensive research into the modification of bitumen.

There are two ways to modify bitumen: one is by the addition of a modifier, the second is to apply a different process during the refining operation.

There are a variety of modifiers used worldwide and these fall into three categories:

- bitumen modified by special processing in the refinery
- plastomeric polymers such as polyethylene (PE) and ethylene vinyl acetate (EVA). These increase the stiffness of the bitumen
- elastomeric polymers such as styrene-butadiene-styrene (SBS). These give the bitumen greater elasticity and cohesive strength

The behaviour of bitumen is such that at low temperatures it behaves as an elastic solid and at high temperatures it behaves as a viscous fluid.

Between these two, it behaves in a visco-elastic manner and as a result of this, the bitumen responds to a wheel load in a similar visco-elastic manner depending upon the wheel load and the temperature.

Whenever a wheel load is applied to a bituminous surface, the resulting strain in the surface firstly gives rise to an elastic response with a gradual increase in the amount of strain until the load is removed, after which time the surface gradually recovers. There is, however, a residual strain which is irrecoverable.

The bitumen is responsible for the behaviour of the flexible surface and it plays a significant part in the behaviour and performance of flexible road pavements and generally, any irrecoverable strains imposed in a flexible pavement increase with loading duration and temperature.

To reduce the occurrence of this problem and to reduce the effects of the irrecoverable strains (i.e. permanent deformation), bitumens are modified, either by stiffening the bitumen to reduce its visco-elastic response, or by increasing its elasticity by reducing its viscosity.

During the refining process, bitumen can be modified by a process known as air rectification or air blowing. Air is blown through the bottom of a distillation column and this reacts with the bitumen to produce a higher viscosity and less temperature-susceptible bitumen.

The addition of modifiers to the bitumen is the other one of the two ways of changing the behaviour of the bitumen. The additives used are thermoplastic rubbers and thermoplastic polymers, of which probably the most commonly used in the UK are EPDM/SBS and EVA respectively.

Thermoplastic polymers increase the viscosity of the bitumen and the deformation resistance of the flexible pavement. Thermoplastic rubbers also increase the viscosity of the bitumen.

Modification of the bitumen during the refining process does not involve the addition of any polymers and the resulting bitumen can be altered to suit the required characteristics at a given temperature. The usual result required is an increase in the viscosity which again results in improved deformation characteristics but also in reduced low temperature stiffness.

Rutting in a flexible pavement occurs when the traffic is channelised and occurs at higher temperatures (e.g. during summer months) when the viscosity of the bitumen is low. The effect of any bitumen modifications therefore is to reduce the risk of permanent deformation at these higher temperatures by increasing its viscosity and reducing the irrecoverable strain.

Bitumen modification is also used to control cracking in the flexible pavement. Traffic loadings, temperature variations (e.g. summer vs winter) and construction methods combine to affect the behaviour of the flexible pavement in use. Cracking occurs in two phases - initiation and propagation - and is referred to as fatigue.

During colder temperatures (e.g. winter months) the bitumen behaves as an elastic solid and this makes the flexible pavement more prone to cracking. The application of bitumen technology means that the effects of these factors which cause fatigue and cracking can be reduced.

12.2 Bitumen testing

There are many types of test which are performed on bitumens, varying from empirical tests which are generally used for specifications through to rheological tests which are used to determine thermoplastic and viscoelastic properties.

From a practical point of view, however, quality assurance (QA) testing at source, e.g. compositional analysis to ensure consistency, and the control of workmanship on site are more relevant.

The standard laboratory tests for bitumen include:

- penetration test
- softening point test (ring and ball method)
- viscosity tests
- Fraas breaking point test

12.2.1 Penetration test

This test measures the consistency (or penetration) of bitumen. In the test, a weighted needle of specific dimensions is allowed to penetrate a sample of bitumen which is maintained at a constant temperature of 25°C. The needle is allowed to penetrate for a duration of five seconds. The distance the needle travels into the bitumen is known as the penetration and is measured in tenths of a millimetre (decimillimetre). The greater the penetration, the less viscous or softer the bitumen.

This test has as its upper and lower limits penetrations of 500 and 2 respectively. Penetrations outside these figures cannot be accurately measured using this test.

The penetration test is a standard test by which standard penetration grade bitumens are classified.

12.2.2 The softening point test (ring and ball)

This test can also be used to determine the consistency of a bitumen. In the test, a steel ball of specific weight is placed on a sample of bitumen contained in a brass ring which is suspended in a bath of either water or glycerine, depending upon the softening point temperature. For higher temperatures (80°C and above), glycerine is used. For lower temperatures (less than 80°C), water is used. The water or glycerine is then heated at a controlled rate to soften the bitumen and allow the ball to pass through the brass ring. When the ball touches a plate set 25mm below the ring, the water/glycerine temperature is recorded. This temperature is known as the softening point.

12.2.3 Viscosity test

The viscosity of a bitumen determines how it will behave at a given temperature or for a given temperature range. There are two measures of viscosity:

- kinematic viscosity. This is measured by the time it takes for a given quantity of bitumen (or any other liquid) to flow through an orifice of a set diameter.
- dynamic viscosity. Also known as absolute viscosity. This is measured by dividing the shear stress applied to the bitumen (or any other liquid) by the rate of shear and is defined as the ratio of shear stress to shear strain.

Capillary and cup viscometers are used to measure the kinematic viscosity of a liquid. In the capillary type, a sample of fluid at a set temperature is allowed to pass through a narrow bore glass tube (or tubes) either under its own weight or or under pressure. The kinematic viscosity is then measured by the time it takes for the bitumen to flow between two points.

A cup viscometer determines the kinematic viscosity of a liquid by allowing a given volume of liquid to pass through an orifice in the base of the cup.

To determine the dynamic viscosity of a liquid, a rotational viscometer is used. These apply a shear stress to the liquid which is maintained at a set temperature (normally above 100°C). The liquid is contained between two surfaces which move relative to each other and measure the resultant shear stresses.

12.2.4 The Fraas breaking point test

This test was developed by Fraas in about 1937 to determine the behaviour of bitumens at low temperatures. It involves the flexing of a thin film of bitumen which is held on a steel plate. The temperature of the plate is gradually lowered until the film cracks. This point is known as the Fraas breaking point.

12.3 Types of bitumen

There are four types of bitumen which are manufactured in the UK to British Standard BS 3690. The standard is divided into three parts to cover road applications, industrial applications and blends of bitumens. The four main types of bitumen are:

- cutback bitumen
- hard bitumen
- oxidised bitumen
- penetration grade bitumen

12.3.1 Cutback bitumen

Cutback bitumens are formed from a blend of kerosene and either 100 pen or 200 pen bitumen. The purpose of the addition of kerosene is to achieve a specific viscosity. Cutback bitumens are used in surface dressing applications and also in standard setting and delayed setting macadams.

12.3.2 Hard and oxidised bitumen

Hard and oxidised bitumens are used predominantly for building and industrial applications such as the manufacture of paints and roofing products and do not generally have a use in flexible pavement construction.

12.3.3 Penetration grade bitumen

Penetration grade bitumens are used predominantly in flexible pavement construction. These bitumens are specified by penetration and softening point and the grades range from 15 pen to 450 pen, with 15 pen being at the hardest end of the range and 450 pen being at the softest. Bitumens graded between 35 pen to 100 pen are normally specified where the stiffness of the bitumen is of significant importance. Softer bitumens from 100 pen to 450 pen tend to be specified where the interlock of the aggregate is more important.

12.4 Evaluation of the performance of bituminous materials

There are many tests currently being proposed for Europe by members of the EU for use in the evaluation of the performance of bituminous materials. Much of this has been brought about by the harmonisation of British and European Standards and a consequent move away from empirical testing methods. A summary of the performance tests used in the evaluation of bituminous materials is given in ***Table 20***.

There are three principal items of testing apparatus which are used in the evaluation of the performance of asphalt and they are used to determine the following five factors:

- fatigue cracking
- durability
- workability
- stiffness
- deformation resistance

The equipment currently in most prevalent use is:

- Nottingham asphalt tester
- gyratory compactor
- wheel tracker

Test for...	Test type
Fatigue cracking	Indirect tensile fatigue test (NAT)
Workability	Gyratory compactor
Stiffness	Indirect tensile stiffness modulus test (NAT)
Deformation resistance	Wheel tracking test

***Table 20** Summary of performance tests for bituminous materials*

12.4.1 Nottingham asphalt tester (NAT)

This piece of equipment is generally used to carry out tests on core samples of bituminous materials, although it has been used on cement stabilised material and lean mix concrete

The test rig consists of a test frame, a pneumatic unit and a data acquisition and control unit which is linked to a computer. In use, the test frame is contained within a temperature controlled environment. There are six tests which this equipment is capable of carrying out:

- indirect tensile fatigue test. This test assesses the resistance of the material to fatigue cracking

- repeated load axial test. This test assesses the resistance to permanent deformation. It gives similar results to the uniaxial creep test but the results are more representative of actual traffic loading conditions
- vacuum repeated load axial test. This test is a further development of the repeated load axial test and again assesses the resistance of a material to permanent deformation
- uniaxial creep. The uniaxial creep test assesses the likely resistance of a material to permanent deformation but is carried out under controlled conditions which may not be an accurate represenation of actual traffic loadings
- indirect tensile stiffness modulus. This test measures the stiffness modulus of a material
- resilient modulus test. This test also measures the stiffness modulus of a material and also the instantaneous and total Poisson's ratio

12.4.2 Gyratory compactor

The gyratory compactor is used to determine the compactability of a sample of asphalt. The sample is heated and placed in a mould and a predetermined static pressure is applied to it. This pressure is maintained as is the mould is gyrated and this causes the aggregate particles in the specimen to move about in a manner similar to that which occurs when the material is compacted under a roller on site.

12.4.3 Wheel tracker

This test is carried out on slabs and sample cores, generally of the wearing course surface. The test apparatus consists of a loaded wheel which applies a load to the test sample, secured on a plate. The plate moves in a backwards and forwards motion beneath the wheel and eventually a rut develops in the sample. The test is carried out for a duration of the shorter of 45 minutes or until a rut of 15mm depth appears.

13

Aspects of pavement construction

13.1 Pavement defects and their treatment

There are three principal defects which can arise in flexibly constructed pavements:

- loss of surface texture
- cracking
- deformation

An outline of the causes and treatment of these defects has been set out in ***Table 21*** and is discussed more fully in the following sections.

13.1.1 Loss of surface texture

Every time a vehicle travels over a road surface, the friction between the vehicle's tyres and the road surface provides the necessary resistance for the vehicle to be propelled in the forward direction. Whenever this occurs, the friction causes abrasion of the aggregate in the wearing course and over a period of time, the aggregates become exposed and polished.

Loss of surface texture also occurs when the wearing course becomes 'fatted up', that is when there is excess binder/filler in relation to the quantity of aggregate. This can be caused by over application of binder, poor quality control and/or workmanship or loss of chippings as a result of trafficking.

Defect		Probable cause	Treatment
Loss of surface texture		Trafficking or poor quality control in application of surface dressing	Resurfacing of pavement
Cracking	Longitudinal cracks in wheelpath	Onset of structural failure	Reconstruction of pavement
	Multiple cracks in wheelpath, general cracking in surface	Onset of structural failure	Reconstruction of pavement
	Longitudinal cracks outside wheelpath	Reflective crack above joint in lower pavement layer	Crack sealing
	Short transverse cracks in surface	Poor construction quality	Crack sealing
	Long transverse cracks in surface	Discontinuity in a lower layer	Reconstruction of pavement if discontinuity is symptom of structural failure of lower layer otherwise crack sealing
Deformation	Narrow ruts	Deformation of upper layers under action of traffic	Resurfacing of pavement
	Wide ruts	Deformation of lower layers of pavement under action of traffic	Reconstruction of pavement if ruts are significant

Table 21 *Defects in flexible pavement construction*

The suitability of the aggregates used in wearing courses, chippings or surface dressing is assessed in the following two tests:

- aggregate abrasion value
- polished stone value

The aggregate abrasion value of an aggregate (or AAV) is a measure of the rate at which chippings or an aggregates wear under the action of traffic. It is similar to the polished stone value (or PSV) of an aggregate which is a measure of the resistance of the chippings or aggregates to polishing.

During wet weather, the resistance to skidding (the friction between the tyre and the road surface) is reduced and it is during wet weather that different road surfaces display the greatest variety of skidding resistance. During dry weather, however, all clean road surfaces offer a high level of skid resistance. Therefore although polished road surfaces present a higher skid risk to vehicles this occurs mainly during wet weather.

The wet weather skidding resistance of a road surface is measured by a Sideway-force Coefficient Routine Investigation Machine, or SCRIM, which travels along a road surface at approximately 50km/h. It is a test that is usually carried out by the highway authorities on public roads.

SCRIM has a test wheel mounted in the middle of the machine in line with the nearside wheel track which rotates in its own plane but which is set at an angle of 20° to the direction of travel. The road surface immediately in front of the test wheel is wetted by a controlled jet of water and as SCRIM moves forward, the test wheel rotates in a forward direction to scuff against the road surface.

This angled rotation of the test wheel and forward movement of SCRIM generates a sideways force and the skidding resistance can be measured as a ratio of the sideways force to the vertical force between the test wheel and the road surface. The results of the SCRIM test have a value which is typically in the range between 0.3 to 0.6.

Methods of repairs to road surfaces which have undergone loss of surface texture include surface dressing and the laying of a new surface either directly on top of the old surface (overlay) or after planing of the existing surface (wearing course replacement). These methods are discussed in subsequent sections of this chapter.

13.1.2 Cracking

The existence of cracks in a road surface is not always indicative of the structural failure of the pavement. It can be caused by a number of factors including quality and control of workmanship during construction, thermal cracking in cement bound lower layers, and reflective cracking above a construction joint.

Table 22 gives an overview of the types of cracks likely to be found in a flexible or flexible composite pavement, together with their possible causes.

If left untreated, cracks permit the ingress of water which will ultimately give rise to deterioration of the road surface especially during winter months as a result of freeze/thaw cycles. Wider cracks may suffer from attrition by heavy vehicles.

13.1.2.1 Reflective cracks

Reflective cracking will occur wherever a flexible surface overlays a joint or other lower layer discontinuity such as a thermal crack. Once a reflection crack has formed, subsequent deterioration of the surface may occur over a relatively short period, depending upon the volume and speed of the traffic.

With reflective cracks, the treatment is generally to cut out the crack for its entire length and to a suitable width using a saw to cut all vertical faces. The vertical sides of the crack may be stepped depending upon the depth of the crack.

The material inside the saw cut is removed and replaced with a suitable material which may incorporate a strip of glass fibre or similar across the width of the crack to provide added resistance to further reflective cracks appearing at the surface.

If the crack is caused by movement of a lower pavement layer such as a rocking concrete slab, then for the repair to be effective it is essential that all movement is eliminated prior to any repairs being undertaken, for example the slab being grouted or re-seated.

Crack description	Possible cause
Long transverse cracks in surface	Discontinuity in a lower pavement layer such as a construction joint, joint in a concrete slab below or a thermal crack in a cement bound roadbase
Short transverse cracks in surface	Poor workmanship during construction
Multiple cracks in wheel track and general surface crazing	Pavement is close to structural failure if pavement is thick (>approximately 200mm) or if cracks are wide. If pavement is thin and cracks are narrow, then pavement may not be close to structural failure
Single cracks in wheel track	Pavement is close to structural failure if pavement is thick (>approximately 200mm), or has a cement bound base. If pavement is thin and cracks are narrow, then pavement may not be close to structural failure
Longitudinal cracks outside wheel track	Reflective crack above a construction joint in a lower layer

Table 22 *Cracking in a flexible pavement*

13.1.2.2 Non-reflective cracks

There are several measures which can be taken to treat cracks in pavements. This section deals with cracks which are not attributable to structural failure which would ordinarily require reconstruction of the pavement. In the treatment of cracks there are six main options:

- slurry sealing and microsurfacing
- overbanding
- cut and patch
- surface dressing
- thin surfacings
- retexturing

13.1.2.3 Slurry sealing and microsurfacing

Slurry sealing and microsurfacing are techniques which can be used effectively to restore and seal crazed surfaces. Both techniques use hand or machine laid mixtures which can contain either conventional or modified binders.

Slurry sealing is generally laid at a depth of 3–5mm and may be used to seal extensively crazed surfaces. Its application is limited to low speed roads because it has a very low texture depth. Prior to application, the surface should be cleaned and any joints and cracks should be raked out or cleaned using compressed air.

The application of the slurry seal is usually from a tanker but in small areas and on footways it may be applied by hand. The materials used to make up the slurry consist of a fine aggregate with filler and water and a selected bitumen emulsion. Trafficking of a slurry sealed surface should be restricted until the material has become stable. Microsurfacing ranges in thickness from approximately 10 to 20mm and is used principally on roads which may be subject to high volumes of traffic.

13.1.2.4 Overbanding

Overbanding is a cheap and efficient way of sealing cracks against the ingress of water, although the application of the seals has several inherent disadvantages.

The first is that overbanding has a much lower skid resistance value and may be up to 3mm proud of the existing surface. Secondly, at night and in the wet, wide overbanding can be mistaken for road markings.

Overbanding should only be used where the cracks will result in an overbanding width of less than 40mm. In certain cases, for example where there is extreme crazing of the surface, it may be more economical to apply a surface dressing or a slurry seal.

Materials used in overbanding applications may include rubberised or filled bitumen compounds and these can provide a textured overbanding surface and are suitable for the treatment of longitudinal joints.

13.1.2.5 Cut and patch

Patching can be used to replace defective areas of pavement when full resurfacing is not considered a feasible option. It is often used where only the wearing course is defective but has also been used extensively for more significant structural pavement repairs in small areas.

Where cracks are such that overbanding is not feasible, e.g. the overbanding would be more than 40mm wide or there would be an excessive amount of overbanding due to crazing of the road surface, then the cracks can be treated by sawing or cutting and patching. If necessary, the crack should be milled out to a depth of 10–20mm or to a depth equal to the wearing course thickness and the edges of the cracks should be saw cut to a vertical face.

In application, the extents of the defective area should be saw cut to form a stable, vertical face and the area of defective pavement within the saw cuts removed and replaced with a suitable mix flexible construction. The saw cuts should form a square or rectangular shape. Prior to laying and compacting any new material within the saw cuts, all vertical faces should be painted with a hot bitumen (50 pen) and the base of the area sprayed with a tack coat. New materials should be laid and compacted in a uniform manner and shaped to the required profile.

After patching the crazed area, the edges of the patch can be overbanded if necessary to prevent the ingress of water and to protect the edges of the repair, or alternatvely for a uniform surface appearance, a larger area of carriageway surface dressed.

13.2 Surface dressing

Surface dressing is the most common of all maintenance treatments and can be used in a variety of locations. The basic principle of surface dressing is to apply a thin layer of bituminous binder to the road surface and to then apply, spread and roll into it a layer of stone chippings. Any cracks should be treated as described in the previous section prior to the surface being dressed. Surface dressing does not give any added strength to the pavement, but instead carries out three other functions:

- it seals the road surface
- it delays road surface deterioration
- it improves skidding resistance

It should be noted that where vehicles perform tight or on-the-spot manouevres or brake sharply, conventional binders used in surface dressing are not able to retain the chippings.

Modified binders should therefore be specified in these instances to increase the life of the surface dressing. There is a simple three stage process which covers surface dressing:

- assessment of design parameters
- selection of aggregates, chippings and binders
- site preparation and traffic management

13.2.1 Assessment of design parameters

In the first instance, four factors need to be considered:

- traffic category
- hardness of old surface
- skidding resistance requirements
- time of laying

13.2.1.1 Traffic category

The traffic category depends on the amount of traffic that the surface is likely to carry and this is defined as being the number of commercial vehicles per day in one direction in the lane under consideration. This may be half the sum in both directions if the former is not known, although where the surface dressing will be applied to a multi-lane carriageway, the specifications for each lane may differ to cater for the higher proportion of commercial vehicles in the left hand (nearside) lane.

13.2.1.2 Hardness of old surface

The hardness of old road surfaces is split into five categories as shown in ***Table 23***. The test to determine the category is a manual one and involves the insertion of a probe into the existing surface when the road surface has a temperature of 15°–35°C. This temperature is then converted to the standard test temperature of 30°C and an equivalent penetration is determined.

13.2.1.3 Skidding resistance requirements

The surface dressing should provide adequate skidding resistance appropriate for the location of the road and the anticipated traffic speed and types. The chippings used in the surface dressing play an important role in this respect. The size, shape, PSV and AAV of the chippings will be determined by the type of surface (hard, soft, etc.) and the amount of traffic using it. For high volumes of traffic, for example, there will be a requirement for a high PSV and AAV.

13.2.1.4 Time of laying

The ambient temperature and road surface temperature can both have an effect on surface dressing and for these reasons, surface dressing is normally carried out from April - September.

Category	Surface description
Very soft	Binder-rich surfaces. Large chippings will easily penetrate into the surface under trafficking action. >12mm penetration by probe.
Soft	Chippings will penetrate easily into the surface under the action of traffic but to a slightly lesser extent than *very soft* category. 8–12mm penetration by probe.
Normal	Chippings will penetrate moderately into the surface under trafficking action. 5–8mm penetration by probe.
Hard	Chippings will penetrate slightly into the surface under trafficking. 2–5 mm penetration by probe
Very hard	Negligible penetration of chippings by trafficking action into surfaces such as concrete. 0–2mm penetration by probe

Table 23 *Classification of existing road surfaces for surface dressing*

13.2.2 Selection of aggregates, chippings and binders

Type of surface, site location and traffic loadings will determine the size and shape of the chippings. These factors will also influence the type of binder to be used and its viscosity. For example, a heavily trafficked road with a soft surface will require a large stone with a high PSV and AAV as well as a more viscous binder than would a lightly trafficked road with a hard surface.

There are three types of chippings used in surface dressing:

- coated chippings
- uncoated chippings
- heated chippings

Coated chippings have a lightly applied coating of binder over their surfaces. They are applied to a surface which also has a binder applied to it and rapid adhesion of the chippings to the binder film is ensured. Uncoated chippings are dry chippings. One disadvantage to the use of uncoated chippings is the fact that a layer of dust is present on the surfaces of the chippings. This can reduce adhesion between the binder film on the surface of the road and the chippings and is a difficulty which is exacerbated at lower temperatures and with smaller chippings. Heated chippings can be either dry or coated. The quantity of chippings applied to the surface is also critical. If too few chippings are applied, then the surface will 'strip'. If there are too many chippings, then this may represent a safety hazard to traffic and may be detrimental to the finished surface.

The selection of binders will be influenced by the location of the site to be surface dressed, the time of year, ambient and road surface temperature at the time of dressing and the anticipated traffic. Binder selection has been simplified and is shown in ***Table 24***.

13.2.3 Site preparation, application and traffic management

All areas of structural weakness should first be repaired, strengthened and filled well in advance of the surface dressing. This is to ensure that the surface to be dressed has as uniform a texture as possible.

Porous surfaces should be sealed with a layer of coated grit or a preliminary dressing of binder and small chippings. Where the surfaces remain porous, these can be sealed using a cement slurry. Any reshaping of the surface should be carried out well in advance of the surface dressing operation when such works involve heating and planing although surface dressing can be carried out immediately after cold-planing operations.

Binder type	Application
Bitumen emulsion	Roads carrying 200–1000 commercial vehicles/day in one direction.
Epoxy resin	Signal controlled junctions, roundabouts, approaches to junctions. Binders modified with bitumen or road tar.
Road tar	Urban sites. Polymer modified binders.
Cut-back bitumen	Concrete roads requiring single surface dressings using natural and synthetic rubber modified binders. Urban sites using polymer modified binders. Heated chippings may be used in urban applications in conjunction with mineral fillers.

Table 24 *Binders for use in surface dressing*

When the binder is applied to the surface, it must be uniformly distributed to ensure even adhesion of the chippings and should not collect in puddles. Prior to applying the binder, the surface should be free from all loose material, dust and other debris. Wherever possible, joints in the binder should not overlap old ones. Further, the safety hazard of joints within the carriageway or wheeltrack is reduced if the joints can be located under white lines.

In a similar manner to the application of the binder, chippings should be applied at a uniform rate. Once applied they should be well rolled in to the surface using pneumatic tyred rollers. The chippings should be rolled in immediately after they have been applied.

The application of surface dressings is of inherently short duration. Once laid, vehicle speeds on a newly dressed surface should be restricted to 20mph to allow sufficient adhesion of the chippings to the surface. In average temperatures and conditions, this may take up to 20 minutes, longer if wet.

Excess chippings should be removed by suction sweeping rather than brushing to minimise the risk of disturbance to the chippings. When ambient temperatures are high, removal of these excess chippings can result in exposure of the binder below and sweeping should immediately cease.

13.2.4 Applications of surface dressing

Surface dressing may be applied to:

- high speed roads
- lower pavement layers
- unbound surfaces
- concrete surfaces
- porous surfaces
- high stress applications

High speed roads may be treated with a surface dressing. The preferred binder is often road tar, but cut-back bitumen and tar–bitumen blends may also be used. Road tar tends to wear and constantly re-expose the aggregate to the effects of abrasion by traffic thus maintaining a better skid resistant surface. Small chippings are preferred (i.e. <14mm nominal size) as this reduces the risk of windscreen damage.

Lower pavement layers such as roadbases may be surface dressed to seal and bind them if they are to be used by construction traffic. Surface dressing of the subgrade or sub-base provides protection against weather damage.

When applying a surface dressing to unbound materials such as gravels, the prime aim is to provide a waterproof surface. The smallest available chipping size should be used in conjunction with a high rate of spread of binder.

Concrete surfaces to be surface dressed using a single dressing should have an aggregate with an AAV of less than 8 and should be no greater than 10mm nominal size. The principal reason for this is that a concrete surface represents the hardest surface type and embedment of the chippings is therefore likely to be minimal. Polymer modified binders are also recommended.

Where concrete surfaces are to be dressed using a double dressing, the first dressing should be applied using 6mm nominal size chippings. The subsequent, second dressing should comprise 10mm chippings used with the same rate of spread of binder as was used for the first dressing.

Surface dressing of porous surfaces should be carried out as for normal surfaces but with the proviso that the chipping size used is one smaller than that which would be normally used on an impervious surface but which is not less than 6mm nominal size.

Where the surface to be dressed is in a high stress situation, such as at busy junctions, crossroads, approaches to traffic lights and roundabouts, then the dressing should incorporate an epoxy resin modified binder, Conventional binders are unable to cope with the stresses imposed by heavy braking and sharp turning and have a reduced ability to retain chippings in these situations. To avoid maintenance difficulties, the current practice is to treat many of these locations with an anti-skid or high friction surface instead of a conventional surface dressing.

13.3 Thin surfacings

There are several systems of thin surfacing currently available. They are generally laid at a thickness of 15–40mm. The application is similar in principle to surface dressing but has the advantage due to its depth that it can be used to regulate minor defects in the running surface such as rutting, when applied over a larger area.

This system also has advantages over wearing course replacement in respect of speed and cost.

13.4 Retexturing

Retexturing can be carried out when there has been a reduction in the skidding resistance of the road surface due either to wear or that bitumen has bled through the surface.

There are several methods of retexturing which include the use of hot compressed air, grit blasting, grinding and bush hammering and these techniques can be used to successfully restore the texture depth and skidding resistance of a smooth surface.

13.5 Milling (planing) and regulating an existing surface

The most common method of removing a carriageway surface is to use a planing machine. Planers come with a variety of cutting widths from about 300mm up to about 2m and can remove up to 150mm surface depth in one pass. Once the surface has been planed, a new surface can be applied, incorporating if necessary a regulating course of bituminous material to even out low or high spots or local depressions.

The machines used can operate with quite a high dimensional tolerance and a great deal of flexibility, so if reshaping or reprofiling of a pavement is required, then it is often more feasible to use a planer than to lay a regulating course of material to achieve the desired falls, particularly when adjacent levels place restrictions on the thickness and therefore finished height of the new surface, for example adjacent to kerbs.

13.6 Overlays

There are four main reasons for overlaying and resurfacing existing pavements as follows:

- to improve ride quality
- to restore skidding resistance
- to replace a defective surface
- to strengthen a pavement

The overlay should be of adequate thickness to resist deformation and cracking and should provide a durable new surface. When overlaying a jointed surface, to resist the appearance of reflection cracks, the incorporation of modified binders and a mesh reinforcement (often referred to as a geogrid) can be used. The use of a thicker overlay will resist reflection cracking more than a thin overlay, but this can be an expensive option if the area to be overlaid is quite considerable.

13.7 Road recycling

Recycling the materials which arise from a pavement during the course of its reconstruction often results in the economic use of readily available materials. Not all recycling processes are suitable for major highways, but they can be used readily on estate and lightly trafficked roads. The recycling process can be carried out either on-site, in which case it is termed in-situ recycling, or off-site, in which case it is termed central plant recycling. Both can be divided into hot and cold processes.

13.7.1 In-situ recycling

The *Repave* process and the remix process are examples of hot in-situ recycling. The *Repave* process involves the bonding of a new surface layer of usually 15–25mm thick centrally recycled material to the existing pavement surface which has been reprofiled, scarified and heated to a depth of about 20mm. The remix process is similar but instead of the surfacing material being centrally recycled, the process utilises the heated and scarified material from the existing road surface. Whichever system is used, the total resulting wearing course thickness should not exceed 60mm.

Both of the above techniques can be used to restore defective pavement surfaces where there are no structural defects. They can also be used for short term repairs in advance of more significant pavement strengthening works.

It should be noted that the *Repave* process is not suited where the pavement surface shows signs of cracking brought about by defective surfacing material. The heating of this material may make it even more defective once it has been laid.

The *Retread* process is an example of in-situ cold recycling. The existing surface is scarified and reshaped then has a new binder added and is recompacted to form a new surface.

13.7.2 Central plant recycling

With central plant recycling, the material is removed from site and processed either for incorporation into a new bituminous mixture (hot recycling) or to be crushed and graded to provide capping material (cold recycling). This method is generally restricted to large scale works.

13.8 Pavement assessment

There are three principal methods of assessing the structural integrity of a pavement:

- high speed road monitor (for trunk roads and motorways)
- visual inspection
- deflection testing

13.8.1 High speed road monitor

The high speed road monitor (HRM) is a fast and effective method of measuring the overall profile and condition of a pavement surface.

It utilises a van and trailer which travel along the pavement surface and which record the longitudinal profile and macrotexture of the nearside wheel track and the horizontal profile of the traffic lane (or approximately the width of the trailer in non-carriageway applications).

To detect the changes in surface profile, lasers are contained within the trailer and feed back information to the van where the information is recorded. The survey can be carried out at speeds of up to 60m.p.h.

13.8.2 Visual inspection

Visual condition surveys are used to identify defects in a pavement and are used in routine assessment of a pavement's structural integrity, often being carried out prior to rehabilitation of an existing road or hard surface. Visual inspections will reveal what mechanical and other surveys cannot easily identify and this is likely to include the following:

- wide cracks
- narrow and hairline cracks
- crazing
- rutting
- overbanding
- small areas of bleeding/fatting up
- staining
- mud pumping

In highway applications, a visual survey usually follows on as a result of other surveys such as HRM.

13.8.3 Deflection testing

Deflection testing is used to assess the structural condition of a pavement by measuring the amount of deflection exhibited by a pavement when a vehicle passes over it. There are two methods used to measure deflection:

- deflection beam
- deflectograph

The results of these tests may be processed in conjunction with traffic data and construction details to give the residual life of the pavement, and overlay requirements to restore a 20 year life.

13.8.3.1 Deflection beam

The deflection beam is a simple manually operated device. It consists of a reference beam located above the pavement surface and fixed to the underside of a heavy goods vehicle and a dial gauge which is located between the twin wheel of the rear axle of the vehicle. As the vehicle travels along the line of the beam, the amount of deflection in the surface is recorded by the dial gauge.

13.8.3.2 Deflectograph

The deflectograph is an advancement of the deflection beam principle. It is a system which is wholly contained within a heavy goods vehicle which travels along a pavement and readings similar to those taken by a deflection beam are recorded at approximately four metre intervals in both wheel tracks. This method is used as a tool for designing overlays and establishing the residual life of the pavement.

13.8.4 Ground penetrating radar

This system operates by transmitting an electromagnetic pulse into the ground. When this technique is applied to a pavement, the different construction layers offer different levels of resistance and the signal passes through them at different velocities. When this occurs, the signal is attenuated and some of it bounces back and the reflected signal is received by a radar receiver which builds up a 'picture' of the interior of the pavement. The system has several limitations in pavement assessment:

- not all pavement details and subsurface features can be identified accurately
- ponding of surface water affects the signal and the system should not be used where there is standing water

Development of this system is in its infancy. It has provided accurate results for the following types of survey to date:

- to detect voids under pavements (e.g. under slabs)
- to detect water under pavements
- to detect changes in construction
- to detect changes in layer thicknesses
- to determine dowel bar alignment

13.9 Surface treatment of rigid pavements

Surfaces of rigid pavements are subject to the same sources of wear as flexibly constructed surfaces. Surfaces can become worn and suffer loss of surface texture in a similar manner and restoration and repair of the surface can be undertaken by any of the four following methods:

- surface dressing
- thin bonded surface repairs
- mechanical roughening
- transverse grooving

13.9.1 Surface dressing

Surface dressing is applied to concrete roads in much the same way as it is to flexible pavements; however, the performance of surface dressing on concrete roads is not always as good as on bituminous roads due to the different coefficients of thermal expansion of concrete and bitumen which can lead to bond failure.

The life span of surface dressings may not be very long, especially where tight manoeuvring by HGVs is anticipated, although this can be overcome to a certain degree by the use of modified binders. Further, the cost of surface dressing small areas can be prohibitively expensive, particularly when modified binders are to be used.

The principal advantages of surface dressing are that it offers a very high low speed skid resistance and restores surface texture.

13.9.2 Thin bonded surface repairs

Thin bonded surface repairs can be used to successfully restore both the microtexture and macrotexture of a concrete road surface. They are relatively expensive and can be time consuming.

The material used is cement mortar for depths up to 20mm and fine concrete for depths over 20mm. Resin mortar is also used, but its use is restricted to small patch repairs or where curing time is critical.

13.9.3 Mechanical roughening

Mechanical roughening of a concrete surface is used to improve skidding resistance. When mechanical roughening is used, the improvements gained are dependent upon the nature of the resulting exposed aggregate. Retexturing of concrete surfaces will bring about an improvement in high speed skidding resistance but at the expense of an increase in noise. Mechanical roughening using blasters, scabblers, grinders and milling equipment can be achieved by one of three methods and can be used to treat large or small areas:

- grit blasting
- bush hammering
- flailing transverse

Grit blasting removes surface polish caused by trafficking action. It offers a short term solution by improving the microtexture of the surface but is only suitable for roads subject to light trafficking. Bush hammering improves the macrotexture of the surface and as with grit blasting removes any surface polish. One disadvantage of this method is that it presents a risk of damage to the surface of the pavement. A flailing transverse improves surface texture and removes surface polish.

13.9.4 Transverse grooving

Transverse grooving can be carried out to restore skidding resistance. It restores braking force and can prevent aquaplaning. It also brings about improvements in surface texture. The process is carried out using diamond cutting discs set into a machine fitted with a grooving head to saw cut irregularly spaced grooves of between 2 and 5mm in depth in strips of over 500mm wide.

13.9.5 Joint repairs

As concrete slabs warm up and cool down, they expand and contract. Joints are provided in slabs during construction to allow for this movement without the slab failing in service. Over time, however, maintenance will be required due to the deterioration of the chemicals used in the seal; this allows the seal to harden, become brittle and break away from the joint. The resulting effect is that debris and moisture can enter the joint causing spalling of the concrete and the formation of sub-surface voids by the washing out of fines in the sub-base. Three classifications of joint repairs are as follows and have been summarised in ***Table 25***:

- compression
- cold applied
- hot applied

13.9.6 Defects in rigid pavements

Four main types of defect can be identified in rigid pavements:

- joint defects
- wear
- cracking
- movement

Joint defects can be caused by poor workmanship during construction, ineffective seals or by uncontrolled movement in the slab.

Classification of jointing material	Predicted longevity	Comments
Compression	Long	Elastomeric seal using polychloropene suitable for all joint types
Cold applied	Medium	Elastomeric seal using polysulphide or polyurethane suitable for all joint types
		Elastomeric seal using silicone suitable for warping joints only
Hot applied	Medium	Elastomeric seal using bitumen, pitch polymer or PVC suitable for all joint types

***Table 25** Joint sealing repairs*

Crack classification	Crack width
Narrow	<0.5mm
Medium	0.5–1.5mm
Wide	>1.5mm

***Table 26** Crack classification*

Crack type	Cause	Repair
Longitudinal crack	Bays constructed too wide Settlement Compression failure Crack inducer missing at joint	Narrow cracks do not require repair unless in an unreinforced slab, in which case a stitched type repair is required
Transverse crack	Bay constructed too long Insufficient lap to reinforcement Joint grooves sawn too late after pouring Dowel bars restrained	Medium cracks in all slab types require a stitch type repair Wide cracks in all slab types require full depth repairs or replacement of whole bay
Transverse cracks at transverse joints	Misalignment of dowels Dowels restrained Joint groove sawn too late after pouring	Full depth repair
Longitudinal cracks at transverse joints	Compression failure	Full depth repair
Longitudinal cracks at longitudinal joints	Mis-aligned crack inducers Omission of crack inducer	Full depth repair Stitched type repair

Table 27 *Cracking in rigid slabs*

Where a joint has been identified as being defective, this is often evidenced by deep spalling at the arrises of the joints and at the corners of bays. Full depth repair of the slab in these situations is generally the only acceptable remedy.

The cause of deep spalling is usually that the dowel bars are restrained or that debris has entered the crack, although the first indications of any movement in the slab may also be noticed as spalling. Wear in rigid pavements can be repaired by resurfacing or retexturing.

Cracking may occur at joints but may also be transverse or longitudinal throughout the slab. Crack classification is given in ***Table 26.*** Cracks are, however, a normal feature of all reinforced slabs. ***Table 27*** gives details of the causes, types and remedies of cracking in rigid pavements.

13.10 Summary of surfacing performance

Table 28 summarises the performance of different types of surface.

	Open textured macadam	Pervious macadam	DBM	Rolled asphalt	Surface Dressing
Deformation resistance	med	med	med/high	med/high	none
Durability	med	low/med	low/med	med/high	med
Noise reduction	med/high	med/high	med	low/med	low
Spray suppression	med	high	low	low/med	med
Riding quality	med/high	med	high	med	low
High speed skid resistance	med/high	high	med/high	med/high	high
Low speed skid resistance	high	high	high	med/high	high

Table 28 *Performance of surfaces*

14

Road design

14.1 General design practice

Road design will generally fall into three categories:

- trunk roads and major highways
- residential/industrial estate roads
- private roads

14.1.1 Trunk roads and major highways schemes

The design of trunk roads and major highways schemes in the UK is carried out to the standards laid down in the Department of the Environment and the Regions (DETR) publication *Design Manual for Roads and Bridges*.

In design, there are two primary considerations. The first is to assess the volume of traffic and the second is to determine the design speed of the road.

The volume of traffic will determine the classification of the road, e.g. motorway, dual carriageway, and the overall pavement construction thickness. For design purposes, it is generally considered that the loads imposed by commercial vehicles contribute far more significantly to the structural damage to the pavement than the loads imposed by private cars; and because of this, the design criteria are based upon the number of commercial vehicles which will use the road. There are many different types of commercial vehicle e.g. articulated lorry, refuse lorry, rigid lorry which each provide different axle loads. In assessing the volume of traffic the different types of commercial vehicles are considered and their overall axle loadings are expressed as an equivalent number of standard axles.

Vehicle speeds determine a number of different factors and the way road users behave relates very much to their perception of the road. The selection of an inappropriate design speed will lead to drivers exceeding the posted speed limits without being aware of the limits of the design. This can lead to high incidents of fatal and serious injury accidents caused by loss of control, overtaking, traffic conflict at junctions and accidents involving pedestrians.

Design speeds should therefore be appropriate to the anticipated speed of the road users, warning signs should be made larger and more conspicuous (high visibility) and the road geometry and features should discourage high vehicle speeds.

The design speed of the road will depend upon the location and the anticipated speed of vehicles using that road and to achieve this, assessment of the 85th percentile vehicle speed should be used. This figure is an estimation of the speed of 85% of the vehicles using that road, i.e. only 15% of the vehicles will exceed that speed. The proposed speed limit of the road is not a reliable design parameter and it should be noted that the design speed will have a significant influence on the road geometry.

The design speed should be selected to correlate the features of the road such as horizontal and vertical geometry (curvature), superelevation and visibility to ensure safe operation of the vehicles which are travelling along it. A guide to the selection of design speeds for different categories of roads is shown in ***Table 29***, although for appropriate design standards, reference should be made to the *Design Manual for Roads and Bridges*.

14.1.2 Residential/industrial estate roads

When residential and industrial estate roads are to be adopted by the highway authority, they will generally be designed to standards set down in design guides issued by the adopting authority. The guides are intended to provide uniformity throughout the area covered by the adopting authority and will have evolved in conjunction with the planning authority to provide a particular streetscape. The guides will usually specify design criteria such as the preferred road layout and geometry, target maximum speeds and traffic calming and may cover, for example, carriageway widths, requirements for footpaths, verges and service margins, horizontal and vertical alignment, maximum and minimum gradients, road construction and pavement thicknesses, distances between junctions, junction radii and suitable/acceptable materials.

The design criteria can be significantly different from those found in the *Design Manual for Roads and Bridges* and will often vary between the areas covered by the adopting authorities.

Speed limit (mph)	Speed limit (kph)	Design speed (kph)	Road type
70	112	120	Motorway
70	112	120	Dual carriageway
60	96	100	Single lane carriageway
50	80	85	Rural carriageway
40	64	70	Urban carriageway
30	48	60	Urban carriageway

Table 29 *Selection of design speeds*

14.1.3 Private roads

Private roads are those that are not maintained by the highway authority and they may be constructed to serve any type of development from a single dwelling house to an industrial estate.

Such roads should be designed to cater for the types of vehicles that are anticipated to use them, i.e. the pavement construction should be of adequate thickness to provide a reasonable design life and the geometry should be such that vehicles can manoeuvre safely and perform turning movements (for example refuse lorries and emergency appliances will require a turning facility at the end of a cul de sac).

Adequate provisions should also be made for drainage and consideration should also be given to the movement of pedestrians and public services if appropriate.

On new estates where the road is to remain in private ownership, it is quite common for the developer to set up a management company to ensure that maintenance of the road is carried out. In this instance, there will often be a regular (e.g. annual) contribution payable by the owners of the road to the management company to achieve this.

In some cases, however, the developer may be requested by the highway authority under the Highways Act 1980 to make a contribution in case the management company should either fail to carry out its functions or should it be deemed that the highway authority could need to make the road up to adopted standards at some point in the future.

14.2 Commonly identified highway design problems

Road safety auditing involves a number of checking processes and assesses, amongst other things, the safety of a road scheme from the perspective of the road user.

For road users at junctions, poor visibility of traffic signal heads has been identified as a recurrent problem. This can lead to 'shunt' type accidents and drivers failing to stop at the signal. Visibility that is good all round can lead to drivers 'seeing through' to the other signals and pedestrians misinterpreting the signals thus stepping into the path of moving traffic. Possible solutions to these problems are to ensure that the signal heads are not obscured by objects such as street furniture, trees, buildings and vegetation and that louvres and cowls are fitted to signals where there is the possibility of driver 'see-through'. Changes in road surface texture and colour can also assist road users in these locations.

At junctions, accidents are relatively common. This is due to the conflicting manouevres performed by opposing streams of traffic. Where drivers are forced to brake suddenly, skidding can be a problem. The application of anti-skid surfaces, retexturing of the road surface and surface dressing treatments can contribute to reducing the risk of skidding.

Safety issues at roundabouts usually arise as a result of high entry speeds and lack of entry path deflection. Vehicles can fail to stop at the give way line and 'shunt' type accidents are a common occurrence, particularly when the lead vehicle brakes suddenly.

Roundabouts are inherently unsafe for the road user on two wheels. One of the principal causes of reported accidents at roundabouts involving two wheeled vehicles is loss of road surface texture. This includes the siting of manhole and chamber covers within the circulatory section of carriageway or on the exit, where there is a greater risk of loss of control due to the higher exit speeds. Features with low skid resistance should be sited out of the circulatory carriageway and exits and high friction treatments could be applied to the entire roundabout.

Solutions to these safety issues include the provision of anti-skid surfacing on the approach to the roundabout, provision of countdown signs and yellow bar markings on the carriageway surface, improvements in the visibility into the roundabout, better signing and as a last resort, increased deflection by road markings and kerb build-outs.

One common safety issue is one of breaks in safety barriers and the inadequate provision of safety fences. This can lead to increases in the severity of the consequences of loss of control accidents, particularly where vehicles descend embankments and collide with fences, trees and other vegetation at the bottom. Street furniture such as lighting columns and traffic signs should wherever possible be sited behind safety fences. Gaps in barriers should be closed to prevent vehicles travelling through the spaces. One final safety point regarding barriers is that at the ends of safety fences, there is the potential for vehicles to be launched off the turned down end of the barrier, particularly at the nosings of slip roads.

Road signs are often obscured by overhanging vegetation. This can cause road users to miss direction and warning signs leading to a failure to react in time to an approaching hazard. Signs should be resited, vegetation trimmed or removed and supplementary signs erected where appropriate.

There are also difficulties for cyclists. Frequently there is inadequate provision made for cyclists, leaving them to mix with other traffic without any protection. Cycle friendly measures are gradually being incorporated into many schemes, but currently the overall view is that such measures are very piecemeal and often leave cyclists stranded between safety features.

Pedestrians also face difficulty, with inadequate provision of dropped kerbs, poor location of crossing points, inadequate provision of safety guardrails and poorly or incorrectly constructed tactile surfaces for people with disabilities.

14.3 General highway design parameters

This section concentrates on urban and rural carriageways designed in accordance with the *Design Manual for Roads and Bridges*. Roads such as new industrial and housing estate roads do not generally fall into this category and are usually subject to the constraints and specifications set down in design guides available from the local highway authority.

Vehicles should be able to make progress along a carriageway safely and without undue hindrance to progress by slower moving vehicles and to achieve this, vertical and horizontal design (the road geometry) should be co-ordinated and should provide adequate, safe overtaking opportunities.

On single carriageway roads, overtaking sections will commence whenever the required forward visibility is achieved either on a straight section of carriageway or on a right hand bend; or where there is sufficient width for the overtaking manouevre to be carried out without encroaching onto the other side of the carriageway. An overtaking section will terminate generally when the required forward visibility is reduced to half.

Before a road can be designed in great detail, there are four factors which need to be considered.

- site topography along the proposed alignment
- site investigation results
- traffic assessment
- options for pavement construction

Once a route has been agreed, the site topography along that route will determine the need for embankments, cuttings, bridges and underpasses. Site investigations will determine how steep any slopes may go and will establish the nature of the existing ground and the level of the water table. From this information, an assessment of the likely ground improvement requirements can be made and also the probable land take requirements which will be governed by the steepness of any embankments or cuttings and the vertical alignment of the road.

The traffic assessment will predict the volume of traffic which will use the road over its design life and from this traffic volume, which is measured in millions of standard axles, the options for pavement construction can be considered.

14.4 Junction types

When designing any road, it is inevitable that there will be a junction of some sort at some point along it. The type of junction can vary depending upon the category of the road, from a simple T-junction to a complex interchange.

For the purposes of this section, there are four main types of junction:

- roundabout
- major/minor junction
- grade-separated junction
- signal controlled junction

14.4.1 Roundabouts

Roundabouts can provide a feasible alternative to grade-separated, at-grade, signal controlled or major/minor priority junctions. The best siting for a roundabout is where the approaches to the roundabout are on a downhill grade. This gives good visibility into each arm of the roundabout, but the engineer should note that a roundabout sited at the end of a long downhill straight may induce high entry speeds and cause drivers to lose control.

The main function of a roundabout is to allow traffic flows to interchange safely and with minimum delay to the road user. In urban areas, traffic movements are likely to have high peak flows and low entry speeds, whereas in rural areas, flows are likely to be less peaked but subject to higher entry speeds.

Driver behaviour may alter as a result of the incorporation of a roundabout, particularly if the construction of the roundabout is intended as an improvement to an existing junction arrangement and especially where the traffic movement out of a minor side road is impeded by large flows on the priority (through) route. Drivers may perceive there is a high risk associated with performing right turn manoeuvres out of such junctions. The incorporation of a roundabout can regularise the flow of traffic and make manoeuvres less hazardous, although at peak periods, signal controls may be required to prevent congestion further downstream.

A second important consideration is that the amount of land required for a roundabout will generally be greater than that required for any of the alternatives, i.e. a signal controlled junction, major/minor junction improvement.

The capacity of a roundabout increases in line with the volume of traffic performing turning movements. Where the proportion of traffic flow through the roundabout to traffic flow turning off the roundabout (i.e. major flow to minor flow) is less than about 3:1, roundabouts provide an economical form of junction.

Safety at roundabouts is influenced by two factors:

- entry deflection
- angle of entry

It is essential, therefore, that vehicles entering the roundabout do so slowly and that vehicles leaving the roundabout do so as quickly as possible Excessive entry speed is probably the main reason for accidents at roundabouts.

In urban and rural locations, normal roundabouts should have an inscribed circle diameter and circulatory carriageway width sufficient to cater for the predicted traffic flows. However, the larger the roundabout, the higher the circulatory speeds and this can cause difficulties in gap assessment which may result in unsafe or awkward manoeuvres especially by elderly and inexperienced drivers and drivers of slow or large vehicles.

Mini roundabouts can be used successfully where it is necessary to provide some form of control over traffic flows.

14.4.2 Major/minor junctions

All major/minor junctions are at-grade junctions. The major road is defined as being the through road onto which the side road (minor road) joins. A major/minor junction can take six different forms:

- simple T-junction
- ghost island junction
- single lane dualling
- crossroads
- skew junction
- left/right staggered junction

A simple T-junction is the simplest form of junction. There are no islands in the major road, although the minor road may have a splitter island or pedestrian refuge.

Ghost island junctions is a development of the simple T-junction. In the major carriageway there will be a ghost island made up from painted lines to direct traffic turning right into the minor road. At each approach to the ghost island, there will be hatched areas on the major carriageway to direct traffic movements.

Single lane dualling is where the right turn into a minor road is protected by a physical island forming a central reservation in the major carriageway. It is an advancement on the ghost island which only has painted islands to direct traffic movements.

A crossroads is where two roads cross at approximately right angles to each other.

A skew junction is a junction where the minor road approaches the major road at an oblique angle.

A staggered junction is a junction of three (or more) roads. The major road continues through and the minor roads connect with it to form consecutive T-junctions on opposite sides of the major road.

Major/minor junctions are the most common form of junction control. Traffic on the minor road gives way to the traffic on the major (through) road. High traffic speeds or significant overtaking manoeuvres in the major road will affect safe egress from the minor road and should be considered during the design stage and if necessary, measures should be taken to reduce vehicle speeds on the major road and to discourage overtaking manoeuvres in the vicinity of the junction.

14.4.3 Grade-separated junctions

Grade-separated junctions are usually constructed as exits and slip roads on and off multi-lane and major carriageways such as trunk roads and motorways. Land take and construction costs for these types of junction can be significantly high if taken in isolation. It should be noted that on low speed single carriageway roads, the use of grade-separated junctions, frequent overbridges and the resulting earthworks can induce higher vehicle speeds by creating the impression of a high speed road. For this reason alone it is not recommended that grade separated junctions are used in locations other than dual carriageways, motorways and where unavoidable due to awkward side road connections.

14.4.4 Signal controlled junctions

Signal controlled junctions may be any of the previously mentioned types of junctions. Traffic signal controls can be installed to regulate traffic flows at almost any form of junction and should be phased so as to ensure smooth and regular traffic patterns.

They can also be installed to assist traffic flows at junctions during peak flow periods as well as providing a permanent control over traffic flows.

14.5 Elements of design

There are four key stages to go through when considering the design of a carriageway:

- select design speed
- determine stopping distances and forward visibility requirements
- determine horizontal geometry
- determine vertical geometry

14.5.1 Stopping distances and forward visibility

Stopping distances are related closely to the design speed and will influence both the horizontal and vertical geometry of the road. In essence, the higher the speed, the greater the required stopping distance.

Requirements for forward visibility around bends will be determined by the stopping distance. This will influence how tight the centreline radius of the road can be or whether widening will be required on the inside of the curve to achieve the necessary forward visibility.

At crests of hills, similar provisions apply. A driver approaching the brow of a hill needs to be able to see what lies beyond and must be able to stop in the distance seen to be clear. The forward visibility will determine how steep each approach to the crest can be and will effectively flatten the severity of the crest.

The engineer should note that there should be no substantial or fixed obstructions to forward visibility within the sight line. Such obstructions can include street furniture such as traffic signs, although slim objects such as lamp columns may be discounted.

14.5.2 Horizontal alignment

The horizontal alignment of the road should be designed to provide:

- safety
- passenger comfort

As a vehicle travels around a radius, the weight is transferred to the outside of the curve. If the radius is too tight for the speed of the vehicle or for the design speed of the road, then the vehicle will fail to negotiate the bend causing a safety hazard, or if it does negotiate the bend, the severity of the curve will cause the occupants of the vehicle some discomfort as they cling to the seat to avoid being thrown out.

To counteract these effects, minimum radii have been set for specific design speeds. For trunk roads, specific centreline radii are set down in the

Design Manual for Roads and Bridges, dependant upon the design speed of the road.

In addition to these minimum radii, there is another factor which can reduce the tendency of the vehicle to want to move to the outside of the curve and this is known as superelevation.

Superelevation is the act of raising the outside edge of the carriageway around the radius so that it is higher than the inside of the curve. This means that the surface of the carriageway falls towards the inside of the radius and as a vehicle travels round the curve, the forces pushing it sideways no longer act wholly in a horizontal direction, but also push the vehicle into the carriageway, causing greater friction between the tyres and the carriageway surface and reducing the likelihood of the vehicle sliding off. Superelevation should be applied progressively up to a maximum of 7%.

Where carriageways are to be constructed less than the standard width, then widening will be required on the inside of the bend so that adequate forward visibility is maintained. The amount of widening varies between 0.3m per lane and 0.6m per lane and is usually applied in urban areas where obstructions mean that substandard radii or carriageway widths are used.

When roads are in cuttings, this requirement for forward visibility becomes more apparent as it is necessary to widen the verges on the inside of the bend so that the embankments do not encroach into the line of sight.

14.5.3 Vertical alignment

The vertical alignment of any carriageway is determined by three factors:

- design speed
- road category
- site topography

The design speed will determine the appropriate parameters in respect of forward visibility particularly at the approaches to summits. The road category will determine the maximum and minimum gradients for the road alignment and the site topography will govern the amount of cut and fill required to construct the road.

In hilly terrain, it is possible to relax the requirements for maximum gradients, but safety should always be the overriding factor. In trunk road deign, if a normal maximum design longitudinal gradient for a road is 6%, but the road passes through a mountain, this gradient may not provide the most economical solution. A relaxation may be permitted but should only be considered alongside safety factors, cost savings and traffic volumes. It should be noted that as road longitudinal gradients rise, there is a progressive decrease in safety. Permitted longitudinal gradients on industrial and residential estate roads are usually specified by the adopting authority.

To effectively drain a carriageway, a minimum longitudinal gradient of 0.5% is recommended. In flat areas, drainage paths can be provided by the incorporation of false channel profiles (local high or low points) or by using over-edge drainage to ditches or French drains. Altering the vertical alignment by the introduction of vertical curves simply to achieve gradients greater than 0.5% is not an economical method of design and should be avoided.

Vertical curves should be provided at all changes in gradient where the algebraic difference in grade is greater than 1%. At crests, the radius of the vertical curve should be of sufficient diameter to allow the full forward visibility distance to be achieved. At sag curves (the opposite of a crest), forward visibility is not a problem during the day. The difficulty arises during hours of darkness when headlamps need to illuminate the road surface. For this reason, sag curves also need to be designed so that they are not too severe.

For the design of vertical curves, the most common method in use is the Aitken and Boyd method which employs several simple formulae to calculate levels and vertical curve length.

In all cases, the approach gradient to the curve is a and the gradient at the other end of the curve is b. When the approach gradient is up and to the right, the gradient is positive; when it is down and to the left, it is negative. The same applies to b.

If the length of the vertical curve is known, the radius can be calculated from the following formula:

$$R = \frac{100L}{(a-b)}$$

where

L = length of curve in metres
R = Radius of curve in metres
a = the approach gradient expressed as a %age
b = the exit gradient expressed as a %age

To find the length of the curve required where the visibility distance is less than the length of the curve use

$$L = \frac{S^2(a-b)}{840}$$

where

L = Length of curve in metres
S = visibility distance in metres
a = the approach gradient expressed as a %age
b = the exit gradient expressed as a %age

To find the length of the curve required where the visibility distance is greater than the length of the curve use

$$L = \frac{2S\,\mathrm{d} - 840}{(a-b)}$$

where

L = Length of curve in metres
S = visibility distance in metres
a = the approach gradient expressed as a %age
b = the exit gradient expressed as a %age

To find the level y at a point x along the vertical curve, the following formula can be used:

$$y = b + \frac{ax}{100} - \frac{x^2}{2R}$$

where

x = the distance from the beginning of the curve to the point under consideration in metres
y = level at point under consideration
b = the level at the beginning of the curve
a = the approach gradient (%)
R = radius of vertical curve

To find the distance to the highest or lowest point x_m along the vertical curve, the following formula can be used:

$$x_m = \left(\frac{a}{(a-b)}\right) \times L$$

where

x_m = distance to lowest point or highest point along curve
a = approach gradient (%)
b = approach gradient (%)
L = length of curve (m)

When wide bridges and verges are incorporated into a single carriageway highway scheme, this can create the false impression of a high speed road and additional signing or other features may be required to overcome this.

Lighting columns particularly on realigned approaches to roundabouts and junctions where the old lighting arrangement has been left untouched can also create a false impression and can lead drivers off the carriageway.

Visibility splays, forward visibility and stopping sight distances are all design parameters that need to be considered during road design. The engineer should be aware of potential obstructions to these such as bridge parapets, earthwork embankments, trees, safety barriers and high 'Trief' type kerbs.

The swept paths of vehicles should also be considered. Major junctions designed in accordance with the *Design Manual for Roads and Bridges* have been developed to cater for 16.5m long articulated vehicles. Overrun areas may be incorporated to avoid unnecessarily large or wide junctions. The swept paths and especially the rear wheel track of long vehicles should be considered so that any overrun areas do not affect the safety of other road users or pedestrians.

Junction layouts should be kept simple. They should also be appropriate for the type of road being constructed and for the environment. A signal controlled junction would not be appropriate on a high speed dual carriageway.

Consideration should also be given to the type of vehicles using the junction and carriageway and the design should reflect any constraints that these vehicles may impose, for example the requirement for crawler lanes, larger than normal entrance/exit radii and overrun areas.

In rural areas, the environmental impact may be significant especially for new rural roads. The engineer should endeavour to minimise this impact and should incorporate and maintain existing features which enhance the appearance of the carriageway and which are more environmentally friendly. Construction methods should also be environmentally friendly.

Traffic management, traffic islands and traffic calming measures should be used where necessary but should not cause unnecessary obstruction or hindrance.

Where traffic speeds are to be kept low, alternative measures to road humps such as tight radii at bends could be incorporated into the design. This is less pertinent for major highway schemes than it is for new industrial and residential estate type developments.

Wide roads in urban areas encourage high speeds, particularly where there is no on-street parking. The incorporation of cycle lanes or central hatched islands with physical islands at regular intervals can reduce vehicle speeds and restrict overtaking thus improving the safety of the road.

Pedestrian guardrailing has five main purposes and should be used where necessary for a specific purpose:

- to highlight to other road users the presence of pedestrian activity
- to prevent pedestrians from spilling into the carriageway
- to direct pedestrians to crossing points
- to discourage pedestrians from crossing the carriageway in undesired or unsafe locations
- to prevent pedestrians from running directly into the carriageway e.g. at school entrances

Skid resistant (also known as anti-skid or high friction) surfaces should be incorporated at approaches to locations where sudden or heavy braking is anticipated from speeds in excess of 20 mph. Such locations will include signal controlled crossings and junctions and roundabouts. It may also be beneficial to apply these surfaces at bends and on single carriageway roads in these locations, the anti-skid surface should extend across the full width of the carriageway so that any vehicles straddling the centreline of the road are not subjected to differential friction thus giving rise to a potential crash risk.

Maintenance of skid resistant surfaces is of significant importance, as under heavy trafficking, these surfaces can wear and become polished relatively quickly.

15

External works and other surfaces

15.1 Introduction

External works can encompass a diverse range of features. This chapter will concentrate on the following:

- bituminous parking areas
- asphalt games and sports areas
- decorative and coloured finishes for asphalt
- high stress applications of asphalt
- types of surfacing
- emergency accesses
- paths, kerbs and channels
- verges, grassed areas and planting

15.2 Bituminous parking areas

Parking areas are effectively flexible pavements on a small scale. They are not generally subjected to the intense loadings of major trunk roads and can therefore be of thinner construction. Difficulties usually arise in car parking areas when the surface is subject to tight on the spot manoeuvres by goods vehicles. ***Table 29*** gives an indication of suitable construction depths for various parking areas.

The engineer should note that the depths shown assume a subgrade with a CBR value of >5%, but should also bear in mind that in light duty applications such as private drives the CBR value has much less significance than in heavy duty applications such as HGV parks.

An assessment of the subgrade should be carried out to determine the overall construction depths particularly where heavier usage is anticipated. As a guide, ***Table 30*** gives an indication of the likely required overall construction thicknesses for different soils

HGV parks	Vehicles carrying up to 5 tonnes payload	Private drives	Public car parks, office car parks
150mm min thickness Type 1 granular material sub-base	150mm min thickness Type 1 granular material sub-base	150mm min thickness clean hardcore	200mm min thickness clean hardcore
150mm thick DBM roadbase 40mm nominal size aggregate	100mm thick DBM roadbase 28mm nominal size aggregate	60mm thick DBM basecourse 20mm nominal size aggregate	60mm thick DBM basecourse 20mm nominal size aggregate
60mm thick DBM basecourse 20mm nominal size aggregate	60mm thick DBM basecourse 20mm nominal size aggregate	30mm thick close graded macadam wearing course 10mm nominal size aggregate	30mm thick close graded macadam wearing course 10mm nominal size aggregate
45mm thick HRA wearing course; mix designation 55/14	30mm thick close graded macadam wearing course; 10mm nominal size aggregate		

Table 29 *Typical flexible car park construction*

For the circulatory areas of car parks where use by HGVs is likely or a regular feature, channelisation should be considered and if necessary the access and circulatory areas designed as a flexible road pavement/carriageway.

	Cohesive soils	Granular soils
Light duty	200–450	200–300
Medium duty	375–600	250–300
Heavy duty	450–650	300–350

Table 30 *Suggested construction thicknesses for parking areas*

15.3 Asphalt games and sports areas

Asphalt in leisure, sport and recreational areas is generally used as final surface layers or as a base for a specialised surface finish. Asphalt surfacings can be porous or non-porous and can be applied to suit the site falls and drainage requirements of the surface. In heavy duty applications, non-porous HRA mixes will be the most durable and are suited for areas such as running tracks or in skateboard parks although mastic asphalt may also be used in the case of the latter. Small areas of surfacing may be hand or machine laid, but more often than not these areas are laid by hand. Softer and more workable mixes are used in hand laying and due to this, when in service during hot temperatures, this material can become soft and easily damaged. Proprietary non-softening mixtures are available but stiffer wearing courses can always be specified.

15.4 Decorative and coloured finishes for asphalt

Asphalt can be finished in a variety of surface colours for a variety of applications such as playgrounds, pedestrian areas and cycle lanes. It can be applied in one of three ways:

- an overall surface treatment can be applied to the wearing course

- the colour can be incorporated into the wearing course mix as a pigment
- coloured decorative chippings can be applied during the laying operation

Surface treatments include surface dressing (refer to Chapter 12) or a pigmented slurry. Pigmented slurries comprise a bitumen emulsion and fine aggregate applied to the surface in a layer of approximately 3mm thickness.

When the colour is to be incorporated into the wearing course, there are three ways this can be achieved:

- a clear resin binder may be used with a coloured aggregate
- a pigment may be added to the asphalt mix
- a coloured aggregate may be used

Coloured decorative chippings may also be applied to the wearing course in a process similar to surface dressing described in Chapter 12. They cannot be rolled into coated macadams or some rolled asphalts.

15.5 High stress applications of asphalt

High stress applications for asphalt occur where there are particularly high loads applied to the pavement, for example in dockyards, airfields, fork lift truck accesses and bus lanes. High stresses also occur where there are PSVs and heavy goods vehicles make turning manoeuvres in tight radii or travel in defined wheel paths.

Many fork lift trucks have steel wheels, make sharp turns and produce high contact pressures. This can also be said of dolly-wheels, cranes and heavy trailers, all of which can produce scuffing and indentation in a normal asphalt surface.

A proper structural assessment of the requirements on the pavements in such high stress applications should be made before any pavement design is undertaken. Information required will include likely wheel loads, traffic movements and soil investigation reports. There are several specialist publications available for the design of pavements in high stress applications, much of which is outside the scope of this book, although the basic principles of pavement design still apply.

15.6 Types of surfacing

There are many types of surface finishes available to a designer. Many of these are standard products available from manufacturers and their outlets. Other materials may be specials and may require manufacture either by machine or by hand.

The properties of these materials vary greatly and it would not be possible to cover all types of surfacing materials and their characteristics in this book. The most commonly encountered surfaces are as follows:

- clay and concrete blocks
- setts, cobbles, bricks and wood blocks
- flags and slabs

15.6.1 Clay and concrete blocks/bricks

Clay and concrete block paving is a common surfacing type used extensively in external works. These materials offer a durable finish with good skid resistant properties, both in the dry and in the wet. In pedestrian areas, the blocks should be laid on a sand bed on a granular sub-base. In areas subject to vehicular traffic, there may be a requirement for a cement bound layer or bituminous roadbase, depending upon the type of traffic anticipated.

15.6.2 Setts, cobbles, bricks and wood blocks

These materials can be used as alternatives to more conventional materials for example in pavement applications, as features, or as a deterrent to pedestrians. It should be noted, however, that in wet weather, these materials offer a significantly low skid resistance.

Cobbles are inherently hard wearing. In lightly trafficked applications, cobbles should be 100mm–250mm nominal size. In pedestrian areas to discourage pedestrian access, cobbles should be slightly larger or set more proud of the adjacent surface. As features for small areas for example adjacent to paths, cobbles should be up to 50mm nominal size; larger areas require 75mm nominal size cobbles generally on economical/quantitative grounds.

Setts are often granite but other materials are used and can include limestone. The setts are generally dressed stone and roughly square or rectangular in shape with a range of standard sizes upwards of 50mm being available.

Conventional bricks may be used in footway surfacing, although it is more common to specify concrete or clay block paving. The bonds used in either brick or block paving are generally similar.

Wood blocks are generally specified for use in pedestrian areas and as features, but can be slippery in the wet especially if subject to algal growth.

15.6.3 Flags and slabs

Standard precast concrete flags and paving slabs are used prevalently throughout urban areas as a means of providing a durable surface for pedestrian and occasional vehicular access. There are two processes used in their manufacture - the hydraulic press and a vibrated open mould. The latter is generally applied when the flags or slabs are to be specials and can often be a more expensive process, particularly in the case of small quantities. Hydraulically pressed items are the most commonly used.

In construction, flags and slabs should be laid on a sand bed on a granular sub-base.

15.7 Paths, kerbs and channels

Paths should be provided wherever there is a need for access. In general, bound footpaths, i.e. those of bituminous construction, are the most commonly found in highway and local authority applications, although other materials such as precast concrete flags, concrete and clay blocks and bricks can also be used.

Footpath crossfalls should usually be between 2% and 5%, depending upon the longitudinal gradient and in adopted footpaths, the crossfalls are generally specified by the adopting authority. The application of a crossfall reduces the risk of puddles and the subsequent likelihood of ice. Steeper crossfalls than 5% can become awkard for pedestrians.

Whatever the material used in the construction, a suitable edge restraint must be provided to prevent the edges of the path from breaking away into the surrounding material.

There are many proprietary systems of restraint currently available which have been specifically designed for a particular surface, but probably the chepaest solution is a timber edging. The timber should be pressure treated timber (e.g. tanalised) and should ideally have 100 x 40mm minimum dimensions and should be with staked using 50mm x 50mm treated timber pegs at 2m minimum centres. Edge restraints for adopted footapths will generally be specified by the adopting authority.

When considering flexible construction, small aggregate size surfacing is generally used, such as 6mm size macadams or fine cold and rolled asphalts. Mastic asphalt can also be used.

Flexible construction is normally two layers of bituminous material with an unbound sub-base. The sub-base is often a general 75mm hardcore for paths in domestic applications, 100mm of Type 1 granular material in areas of high pedestrian traffic or 150mm of Type 1 granular material in areas subjet to occasional vehicular access (e.g. where vehicles cross over a footpath into a driveway).

The flexible construction is usually no more than 60–65mm, often as a 40mm thick basecourse with 25mm thick wearing course.

Current practice for longitudinal gradients is no steeper than 1:12 with a preference for a maximum of 1:15–1:20 wherever possible.

In use, kerbs perform three main functions:

- to define and support a footpath or carriageway
- to provide a postive drainage channel to direct surface water run-off

- to provide a suitable edge restraint

There are many types of kerbs available which will more than adequately perform these functions and materials may include stone (rough, hewn or cast), precast or in-situ concrete and proprietary manufactured items.

The kerbs themselves should be set in a minimum bed and haunch of 100mm thick concrete. Where the kerb is adjacent to a carriageway this should be increased to 150mm in order to resist the forces from wheel impact and loadings.

15.8 Emergency accesses

Emergency access will be required to the perimeters of some buildings and will need to be capable of withstanding the loadings imposed by emergency vehicles such as fire tenders. These accesses may also be used for refuse collection and for access by maintenance vehicles.

The materials used will generally depend upon the employer's requirements for the project or the agreed specification. Unbound materials such as gravels or hoggin may be used but should be angular rather than rounded to allow for adequate compaction. Self-setting unbound materials are also available such as limestone gravels of specific gradings.

Bound materials may be used, as may concrete and other surfacings such as hollow or perforated blocks which retain topsoil and can be seeded to allow vegetation to grow.

It should be noted that although emergency accesses tend to be used very infrequently and can be constructed at very little relative cost, there is little point in designing an access with such weak construction that repairs are required immediately after the first fire engine has performed a turning manoeuvre on it.

For example, if a Type 1 granular material is to be used, a nominal thickness of 225mm would normally be adequate assuming a subgrade CBR of 2%. Anything less than this (except on stiffer subgrades) is likely to require strengthening, for example by the incorporation of a geotextile to prevent the aggregates from disappearing into the subgrade.

15.9 Verges, grassed areas and planting

Verges and grassed areas should be considered as integral design elements and features. Narrow verges and strips of landscaping may represent maintenance difficulties, particularly if access is awkward or there is a high volume of pedestrian traffic. The primary function of landscaping is one of appearance, although there are two secondary functions:

- landscaping can be used to absorb noise
- planting and landscaping can be used to prevent soil erosion and to stabilise the soil

In addition to the above, landscaping and planting are suited to the coverage of large areas because they can be relatively inexpensive to construct, is easy to maintain and can be readily shaped to suit the available materials or other features. Grass seeds should be selected for the proposed location and different mixes will be required, for example, in verges, slopes, central reserves or prestigious areas.

When planning verges and grassed areas, consideration should be given to the volumes of pedestrian traffic over the area; if they are excessive, then either alternative surfaces may need to be considered or at least a variety of accesses to prevent pedestrian flows from becoming channelised.

For ease of maintenance, gradients generally should be no steeper than 1:3, although in highway embankments, slopes steeper than this may be unavoidable. Verge widths should not be too narrow. It is preferable to have one wide verge on one side of a carriageway only than two very narrow ones, one per side.

Index